Nanocomposites: Fundamentals, Technology and Applications

Nanocomposites: Fundamentals, Technology and Applications

Edited by
Emmanuel Craig

Larsen & Keller
www.larsen-keller.com

Nanocomposites: Fundamentals, Technology and Applications
Edited by Emmanuel Craig
ISBN: 978-1-63549-192-0 (Hardback)

☰ Larsen & Keller

Published by Larsen and Keller Education,
5 Penn Plaza,
19th Floor,
New York, NY 10001, USA

Cataloging-in-Publication Data

Nanocomposites : fundamentals, technology and applications / edited by Emmanuel Craig.
    p. cm.
Includes bibliographical references and index.
ISBN 978-1-63549-192-0
1. Nanocomposites (Materials). 2. Composite materials--Technological innovations. 3. Nanostructured materials.
I. Craig, Emmanuel.
TA418.9.N35 N36 2017
620.5--dc23

The publisher's policy is to use permanent paper from mills that operate a sustainable forestry policy. Furthermore, the publisher ensures that the text paper and cover boards used have met acceptable environmental accreditation standards.

Printed and bound in the United States of America.

For more information regarding Larsen and Keller Education and its products, please visit the publisher's website www.larsen-keller.com

Table of Contents

Preface

This book attempts to understand the multiple branches that fall under the discipline of nanocomposites and how such concepts have practical applications. Nanocomposites refer to the material with two to three dimensions measuring less than 100 nanometers. The different types of nanocomposites are metal-matrix nanocomposites, polymer-matrix nanocomposites and ceramic-matrix nanocomposites, etc. The various sub-fields of nanocomposites along with technological progress that have future implications are glanced at in this book. Most of the topics introduced in this book cover new techniques and the applications of nanocomposites. Different approaches, evaluations and methodologies and advanced studies have been included in it. This textbook is meant for students who are looking for a detailed explanation of the technologies related to nanocomposites.

A short introduction to every chapter is written below to provide an overview of the content of the book:

Chapter 1 - Nanocomposites are described as multiphase solid materials where at least one of the phases has one or two dimensions less than 100 nanometers or structures. In this chapter, the reader is introduced to nanocomposites and an overview of natural and man-made nanocomposites; **Chapter 2** - This chapter introduces the concept of nanoparticles and traces the history of their use. The chapter discusses the properties, functionality, synthesis, morphology and characterization of nanoparticles. There is a section of this chapter dedicated to the specialized nanoparticle–biomolecule conjugate, which is a nanoparticle with biomolecules attached to its surface; **Chapter 3** - Nanocomposites can be classified under metal matrix-based nanocomposites, polymer matrix nanocomposites and ceramic matrix nanocomposites. This chapter provides a detailed account of the varying chemical, structural, thermal and electrical conductivities and magnetic susceptibilities of the various types of nanocomposites. The chapter also deals with topics like carbon nanotube, silver nanoparticle, hybrid material, nanowire and nanocellulose; **Chapter 4** - Nanocomposites find application in various fields and there are several technologies that singularly depend on nanocomposites. This chapter lists and describes technologies like titanate nanosheet, ceramic engineering and polyethylene terephthalate. The topics discussed in the chapter are of great importance to broaden the existing knowledge on nanocomposites; **Chapter 5** - Nanocomposites are used in a variety of devices and technologies that provide more efficiency and longer-life. Some of these applications that have been listed in this chapter are light-emitting diode, photodiode, solar film, colloid, gel etc. Nanocomposites are best understood in confluence with the major topics listed in the following chapter.

I extend my sincere thanks to the publisher for considering me worthy of this task. Finally, I thank my family for being a source of support and help.

Editor

Introduction to Nanocomposites

Nanocomposites are described as multiphase solid materials where at least one of the phases has one or two dimensions less than 100 nanometers or structures. In this chapter, the reader is introduced to nanocomposites and an overview of natural and man-made nanocomposites.

Nanocomposite is a multiphase solid material where one of the phases has one, two or three dimensions of less than 100 nanometers (nm), or structures having nano-scale repeat distances between the different phases that make up the material. In the broadest sense this definition can include porous media, colloids, gels and copolymers, but is more usually taken to mean the solid combination of a bulk matrix and nano-dimensional phase(s) differing in properties due to dissimilarities in structure and chemistry. The mechanical, electrical, thermal, optical, electrochemical, catalytic properties of the nanocomposite will differ markedly from that of the component materials. Size limits for these effects have been proposed, <5 nm for catalytic activity, <20 nm for making a hard magnetic material soft, <50 nm for refractive index changes, and <100 nm for achieving superparamagnetism, mechanical strengthening or restricting matrix dislocation movement.

Nanocomposites are found in nature, for example in the structure of the abalone shell and bone. The use of nanoparticle-rich materials long predates the understanding of the physical and chemical nature of these materials. Jose-Yacaman *et al.* investigated the origin of the depth of colour and the resistance to acids and bio-corrosion of Maya blue paint, attributing it to a nanoparticle mechanism. From the mid-1950s nanoscale organo-clays have been used to control flow of polymer solutions (e.g. as paint viscosifiers) or the constitution of gels (e.g. as a thickening substance in cosmetics, keeping the preparations in homogeneous form). By the 1970s polymer/clay composites were the topic of textbooks, although the term "nanocomposites" was not in common use.

In mechanical terms, nanocomposites differ from conventional composite materials due to the exceptionally high surface to volume ratio of the reinforcing phase and/or its exceptionally high aspect ratio. The reinforcing material can be made up of particles (e.g. minerals), sheets (e.g. exfoliated clay stacks) or fibres (e.g. carbon nanotubes or electrospun fibres). The area of the interface between the matrix and reinforcement phase(s) is typically an order of magnitude greater than for conventional composite materials. The matrix material properties are significantly affected in the vicinity of the reinforcement. Ajayan *et al.* note that with polymer nanocomposites, properties related to local chemistry, degree of thermoset cure, polymer chain mobility, polymer chain conformation, degree of polymer chain ordering or crystallinity can all vary significantly and continuously from the interface with the reinforcement into the bulk of the matrix.

This large amount of reinforcement surface area means that a relatively small amount of nanoscale reinforcement can have an observable effect on the macroscale properties of the composite. For example, adding carbon nanotubes improves the electrical and thermal conductivity. Other kinds of nanoparticulates may result in enhanced optical properties, dielectric properties, heat resistance or mechanical properties such as stiffness, strength and resistance to wear and damage. In general, the nano reinforcement is dispersed into the matrix during processing. The percentage by weight (called *mass fraction*) of the nanoparticulates introduced can remain very low (on the order of 0.5% to 5%) due to the low filler percolation threshold, especially for the most commonly used non-spherical, high aspect ratio fillers (e.g. nanometer-thin platelets, such as clays, or nanometer-diameter cylinders, such as carbon nanotubes). The orientation and arrangement of asymmetric nanoparticles, thermal property mismatch at the interface, interface density per unit volume of nanocomposite, and polydispersity of nanoparticles significantly affect the effective thermal conductivity of nanocomposites.

Ceramic-matrix Nanocomposites

In this group of composites the main part of the volume is occupied by a ceramic, i.e. a chemical compound from the group of oxides, nitrides, borides, silicides etc.. In most cases, ceramic-matrix nanocomposites encompass a metal as the second component. Ideally both components, the metallic one and the ceramic one, are finely dispersed in each other in order to elicit the particular nanoscopic properties. Nanocomposite from these combinations were demonstrated in improving their optical, electrical and magnetic properties as well as tribological, corrosion-resistance and other protective properties.

The binary phase diagram of the mixture should be considered in designing ceramic-metal nanocomposites and measures have to be taken to avoid a chemical reaction between both components. The last point mainly is of importance for the metallic component that may easily react with the ceramic and thereby lose its metallic character. This is not an easily obeyed constraint, because the preparation of the ceramic component generally requires high process temperatures. The most safe measure thus is to carefully choose immiscible metal and ceramic phases. A good example for such a combination is represented by the ceramic-metal composite of TiO_2 and Cu, the mixtures of which were found immiscible over large areas in the Gibbs' triangle of Cu-O-Ti.

The concept of ceramic-matrix nanocomposites was also applied to thin films that are solid layers of a few nm to some tens of μm thickness deposited upon an underlying substrate and that play an important role in the functionalization of technical surfaces. Gas flow sputtering by the hollow cathode technique turned out as a rather effective technique for the preparation of nanocomposite layers. The process operates as a vacuum-based deposition technique and is associated with high deposition rates up to some μm/s and the growth of nanoparticles in the gas phase. Nanocomposite layers in the ceramics range of composition were prepared from TiO_2 and Cu by the hollow cathode technique that showed a high mechanical hardness, small coefficients of friction and a high resistance to corrosion.

Metal-matrix Nanocomposites

Metal matrix nanocomposites can also be defined as reinforced metal matrix composites. This type of composites can be classified as continuous and non-continuous reinforced materials. One of the more important nanocomposites is Carbon nanotube metal matrix composites, which is an emerging new material that is being developed to take advantage of the high tensile strength and electrical conductivity of carbon nanotube materials. Critical to the realization of CNT-MMC possessing optimal properties in these areas are the development of synthetic techniques that are

(a) economically producible, (b) provide for a homogeneous dispersion of nanotubes in the metallic matrix, and (c) lead to strong interfacial adhesion between the metallic matrix and the carbon nanotubes. In addition to carbon nanotube metal matrix composites, boron nitride reinforced metal matrix composites and carbon nitride metal matrix composites are the new research areas on metal matrix nanocomposites.

A recent study, comparing the mechanical properties (Young's modulus, compressive yield strength, flexural modulus and flexural yield strength) of single- and multi-walled reinforced polymeric (polypropylene fumarate—PPF) nanocomposites to tungsten disulfide nanotubes reinforced PPF nanocomposites suggest that tungsten disulfide nanotubes reinforced PPF nanocomposites possess significantly higher mechanical properties and tungsten disulfide nanotubes are better reinforcing agents than carbon nanotubes. Increases in the mechanical properties can be attributed to a uniform dispersion of inorganic nanotubes in the polymer matrix (compared to carbon nanotubes that exist as micron sized aggregates) and increased crosslinking density of the polymer in the presence of tungsten disulfide nanotubes (increase in crosslinking density leads to an increase in the mechanical properties). These results suggest that inorganic nanomaterials, in general, may be better reinforcing agents compared to carbon nanotubes.

Another kind of nanocomposite is the energetic nanocomposite, generally as a hybrid sol–gel with a silica base, which, when combined with metal oxides and nano-scale aluminum powder, can form *superthermite* materials.

Polymer-matrix Nanocomposites

In the simplest case, appropriately adding nanoparticulates to a polymer matrix can enhance its performance, often dramatically, by simply capitalizing on the nature and properties of the nanoscale filler (these materials are better described by the term *nanofilled polymer composites*). This strategy is particularly effective in yielding high performance composites, when good dispersion of the filler is achieved and the properties of the nanoscale filler are substantially different or better than those of the matrix.

Nanoparticles such as graphene, carbon nanotubes, molybdenum disulfide and tungsten disulfide are being used as reinforcing agents to fabricate mechanically strong biodegradable polymeric nanocomposites for bone tissue engineering applications. The addition of these nanoparticles in the polymer matrix at low concentrations (~0.2 weight %) cause significant improvements in the compressive and flexural mechanical properties of polymeric nanocomposites. Potentially, these nanocomposites may be used as a novel, mechanically strong, light weight composite as bone implants. The results suggest that mechanical reinforcement is dependent on the nanostructure morphology, defects, dispersion of nanomaterials in the polymer matrix, and the cross-linking density of the polymer. In general, two-dimensional nanostructures can reinforce the polymer better than one-dimensional nanostructures, and inorganic nanomaterials are better reinforcing agents than carbon based nanomaterials. In addition to mechanical properties, polymer nanocomposites based on carbon nanotubes or graphene have been used to enhance a wide range of properties, giving rise to functional materials for a wide range of high added value applications in fields such as energy conversion and storage, sensing and biomedical tissue engineering. For example, multi-walled carbon nanotubes based polymer nanocomposites have been used for the enhancement of the electrical conductivity.

Nanoscale dispersion of filler or controlled nanostructures in the composite can introduce new physical properties and novel behaviors that are absent in the unfilled matrices. This effectively changes the nature of the original matrix (such composite materials can be better described by the term *genuine nanocomposites or hybrids*). Some examples of such new properties are fire resistance or flame retardancy, and accelerated biodegradability.

A range of polymeric nanocomposites are used for biomedical applications such as tissue engineering, drug delivery, cellular therapies. Due to unique interactions between polymer and nanoparticles, a range of property combinations can be engineered to mimic native tissue structure and properties. A range of natural and synthetic polymers are used to design polymeric nanocomposites for biomedical applications including starch, cellulose, alginate, chitosan, collagen, gelatin, and fibrin, poly(vinyl alcohol) (PVA), poly(ethylene glycol) (PEG), poly(caprolactone) (PCL), poly(lactic-co-glycolic acid) (PLGA), and poly(glycerol sebacate) (PGS). A range of nanoparticles including ceramic, polymeric, metal oxide and carbon-based nanomaterials are incorporated within polymeric network to obtain desired property combinations.

References

- Gatti, Teresa; Vicentini, Nicola; Mba, Miriam; Menna, Enzo (2016-02-01). "Organic Functionalized Carbon Nanostructures for Functional Polymer-Based Nanocomposites". European Journal of Organic Chemistry. 2016 (6): 1071–1090. doi:10.1002/ejoc.201501411. ISSN 1099-0690.

- Janeta, Mateusz; John, Łukasz; Ejfler, Jolanta; Szafert, Sławomir (2014-11-24). "High-Yield Synthesis of Amido-Functionalized Polyoctahedral Oligomeric Silsesquioxanes by Using Acyl Chlorides". Chemistry: A European Journal. 20 (48): 15966–15974. doi:10.1002/chem.201404153. ISSN 1521-3765.

- Lalwani, Gaurav; Henslee, Allan M.; Farshid, Behzad; Lin, Liangjun; Kasper, F. Kurtis; Yi-, Yi-Xian; Qin, Xian; Mikos, Antonios G.; Sitharaman, Balaji (2013). "Two-dimensional nanostructure-reinforced biodegradable polymeric nanocomposites for bone tissue engineering". Biomacromolecules. 14 (3): 900–909. doi:10.1021/bm301995s. PMC 3601907. PMID 23405887.

- Lalwani, Gaurav; Henslee, A. M.; Farshid, B; Parmar, P; Lin, L; Qin, Y. X.; Kasper, F. K.; Mikos, A. G.; Sitharaman, B (September 2013). "Tungsten disulfide nanotubes reinforced biodegradable polymers for bone tissue engineering". Acta Biomaterialia. 9 (9): 8365–8373. doi:10.1016/j.actbio.2013.05.018. PMC 3732565. PMID 23727293.

- Tian, Zhiting; Hu, Han; Sun, Ying (2013). "A molecular dynamics study of effective thermal conductivity in nanocomposites". Int. J. Heat Mass Transfer. 61: 577–582. doi:10.1016/j.ijheatmasstransfer.2013.02.023.

- Lalwani, G; Henslee, AM; Farshid, B; Parmar, P; Lin, L; Qin, YX; Kasper, FK; Mikos, AG; Sitharaman, B (September 2013). "Tungsten disulfide nanotubes reinforced biodegradable polymers for bone tissue engineering". Acta biomaterialia. 9 (9): 8365–73. doi:10.1016/j.actbio.2013.05.018. PMC 3732565. PMID 23727293.

2

Nanoparticles: An Overview

This chapter introduces the concept of nanoparticles and traces the history of their use. The chapter discusses the properties, functionality, synthesis, morphology and characterization of nanoparticles. There is a section of this chapter dedicated to the specialized nanoparticle–biomolecule conjugate, which is a nanoparticle with biomolecules attached to its surface.

Nanoparticle

Nanoparticles are particles between 1 and 100 nanometers in size. In nanotechnology, a particle is defined as a small object that behaves as a whole unit with respect to its transport and properties. Particles are further classified according to diameter. Ultrafine particles are the same as nanoparticles and between 1 and 100 nanometers in size, fine particles are sized between 100 and 2,500 nanometers, and coarse particles cover a range between 2,500 and 10,000 nanometers. Scientific research on nanoparticles is intense as they have many potential applications in medicine, optics, and electronics. The U.S. National Nanotechnology Initiative offers government funding focused on nanoparticle research.

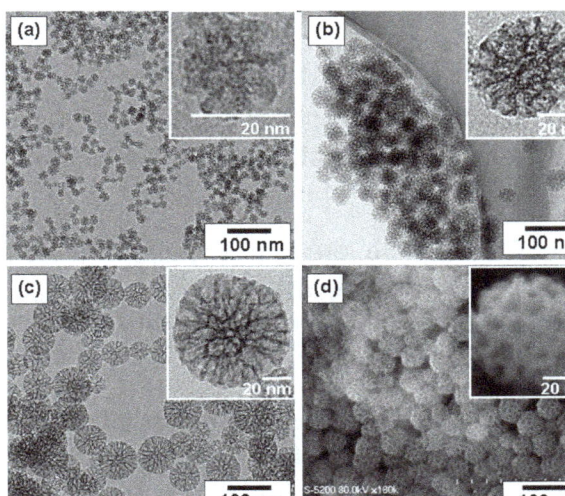

TEM (a, b, and c) images of prepared mesoporous silica nanoparticles with mean outer diameter: (a) 20nm, (b) 45nm, and (c) 80nm. SEM (d) image corresponding to (b). The insets are a high magnification of mesoporous silica particle.

Definition

IUPAC Definition

Particle of any shape with dimensions in the 1×10^{-9} and 1×10^{-7} m range.

Note 1: Modified from definitions of nanoparticle and nanogel in [refs.,].

Note 2: The basis of the 100-nm limit is the fact that novel properties that differentiate particles from the bulk *material* typically develop at a critical length scale of under 100 nm.

Note 3: Because other phenomena (transparency or turbidity, ultrafiltration, stable dispersion, etc.) that extend the upper limit are occasionally considered, the use of the prefix *nano* is accepted for dimensions smaller than 500 nm, provided reference to the definition is indicated.

Note 4: Tubes and fibers with only two dimensions below 100 nm are also nanoparticles.

The term "nanoparticle" is not usually applied to individual molecules; it usually refers to inorganic materials.

The reason for the synonymous definition of nanoparticles and ultrafine particles is that, during the 1970s and 80s, when the first thorough fundamental studies with "nanoparticles" were underway in the USA (by Granqvist and Buhrman) and Japan, (within an ERATO Project) they were called "ultrafine particles" (UFP). However, during the 1990s before the National Nanotechnology Initiative was launched in the USA, the new name, "nanoparticle," had become more common (for example, see the same senior author's paper 20 years later addressing the same issue, lognormal distribution of sizes). Nanoparticles can exhibit size-related properties significantly different from those of either fine particles or bulk materials.

Nanoclusters have at least one dimension between 1 and 10 nanometers and a narrow size distribution. Nanopowders are agglomerates of ultrafine particles, nanoparticles, or nanoclusters. Nanometer-sized single crystals, or single-domain ultrafine particles, are often referred to as nanocrystals.

Background

Although nanoparticles are associated with modern science, they have a long history. Nanoparticles were used by artisans as far back as the ninth century in Mesopotamia for creating a glittering effect on the surface of pots.

In modern times, pottery from the Middle Ages and Renaissance often retains a distinct gold- or copper-colored metallic glitter. This luster is caused by a metallic film that was applied to the transparent surface of a glazing. The luster can still be visible if the film has resisted atmospheric oxidation and other weathering.

The luster originates within the film itself, which contains silver and copper nanoparticles dispersed homogeneously in the glassy matrix of the ceramic glaze. These nanoparticles are created by the artisans by adding copper and silver salts and oxides together with vinegar, ochre, and clay on the surface of previously-glazed pottery. The object is then placed into a kiln and heated to about 600 °C in a reducing atmosphere.

In the heat the glaze softens, causing the copper and silver ions to migrate into the outer layers of the glaze. There the reducing atmosphere reduced the ions back to metals, which then came together forming the nanoparticles that give the color and optical effects.

Luster technique showed that ancient craftsmen had a sophisticated empirical knowledge of materials. The technique originated in the Muslim world. As Muslims were not allowed to use gold in artistic representations, they sought a way to create a similar effect without using real gold. The solution they found was using luster.

Michael Faraday provided the first description, in scientific terms, of the optical properties of nanometer-scale metals in his classic 1857 paper. In a subsequent paper, the author (Turner) points out that: "It is well known that when thin leaves of gold or silver are mounted upon glass and heated to a temperature that is well below a red heat (~500 °C), a remarkable change of properties takes place, whereby the continuity of the metallic film is destroyed. The result is that white light is now freely transmitted, reflection is correspondingly diminished, while the electrical resistivity is enormously increased."

Uniformity

The chemical processing and synthesis of high-performance technological components for the private, industrial, and military sectors requires the use of high-purity ceramics, polymers, glass-ceramics, and composite materials. In condensed bodies formed from fine powders, the irregular particle sizes and shapes in a typical powder often lead to non-uniform packing morphologies that result in packing density variations in the powder compact.

Uncontrolled agglomeration of powders due to attractive van der Waals forces can also give rise to in microstructural heterogeneity. Differential stresses that develop as a result of non-uniform drying shrinkage are directly related to the rate at which the solvent can be removed, and thus highly dependent upon the distribution of porosity. Such stresses have been associated with a plastic-to-brittle transition in consolidated bodies, and can yield to crack propagation in the unfired body if not relieved.

In addition, any fluctuations in packing density in the compact as it is prepared for the kiln are often amplified during the sintering process, yielding inhomogeneous densification. Some pores and other structural defects associated with density variations have been shown to play a detrimental role in the sintering process by growing and thus limiting end-point densities. Differential stresses arising from inhomogeneous densification have also been shown to result in the propagation of internal cracks, thus becoming the strength-controlling flaws.

Inert gas evaporation and inert gas deposition are free many of these defects due to the distillation (cf. purification) nature of the process and having enough time to form single crystal particles, however even their non-aggreated deposits have lognormal size distribution, which is typical with nanoparticles. The reason why modern gas evaporation techniques can produce a relatively narrow size distribution is that aggregation can be avoided. However, even in this case, random residence times in the growth zone, due to the combination of drift and diffusion, result in a size distribution appearing lognormal.

It would, therefore, appear desirable to process a material in such a way that it is physically uniform with regard to the distribution of components and porosity, rather than using particle size distributions that will maximize the green density. The containment of a uniformly dispersed assembly of strongly interacting particles in suspension requires total control over interparticle forces. Monodisperse nanoparticles and colloids provide this potential.

Monodisperse powders of colloidal silica, for example, may therefore be stabilized sufficiently to ensure a high degree of order in the colloidal crystal or polycrystalline colloidal solid that results from aggregation. The degree of order appears to be limited by the time and space allowed for longer-range correlations to be established. Such defective polycrystalline colloidal structures would appear to be the basic elements of submicrometer colloidal materials science and, therefore, provide the first step in developing a more rigorous understanding of the mechanisms involved in microstructural evolution in high performance materials and components.

Properties

Nanoparticles are of great scientific interest as they are, in effect, a bridge between bulk materials and atomic or molecular structures. A bulk material should have constant physical properties regardless of its size, but at the nano-scale size-dependent properties are often observed. Thus, the properties of materials change as their size approaches the nanoscale and as the percentage of the surface in relation to the percentage of the volume of a material becomes significant. For bulk materials larger than one micrometer (or micron), the percentage of the surface is insignificant in relation to the volume in the bulk of the material. *The interesting and sometimes unexpected properties of nanoparticles are therefore largely due to the large surface area of the material, which dominates the contributions made by the small bulk of the material.*

Silicon nanopowder

Nanoparticles often possess unexpected optical properties as they are small enough to confine their electrons and produce quantum effects. For example, gold nanoparticles appear deep-red to black in solution. Nanoparticles of yellow gold and grey silicon are red in color. Gold nanoparticles melt at much lower temperatures (~300 °C for 2.5 nm size) than the gold slabs (1064 °C);. Absorption of solar radiation is much higher in materials composed of nanoparticles than it is in thin films of continuous sheets of material. In both solar PV and solar thermal applications, controlling the size, shape, and material of the particles, it is possible to control solar absorption.

1 kg of particles of 1 mm³ has the same surface area as 1 mg of particles of 1 nm³

Other size-dependent property changes include quantum confinement in semiconductor particles, surface plasmon resonance in some metal particles and superparamagnetism in magnetic materials. What would appear ironic is that the changes in physical properties are not always desirable. Ferromagnetic materials smaller than 10 nm can switch their magnetisation direction using room temperature thermal energy, thus making them unsuitable for memory storage.

Suspensions of nanoparticles are possible since the interaction of the particle surface with the solvent is strong enough to overcome density differences, which otherwise usually result in a material either sinking or floating in a liquid.

The high surface area to volume ratio of nanoparticles provides a tremendous driving force for diffusion, especially at elevated temperatures. Sintering can take place at lower temperatures, over shorter time scales than for larger particles. In theory, this does not affect the density of the final product, though flow difficulties and the tendency of nanoparticles to agglomerate complicates matters. Moreover, nanoparticles have been found to impart some extra properties to various day to day products. For example, the presence of titanium dioxide nanoparticles imparts what we call the self-cleaning effect, and, the size being nano-range, the particles cannot be observed. Zinc oxide particles have been found to have superior UV blocking properties compared to its bulk substitute. This is one of the reasons why it is often used in the preparation of sunscreen lotions, is completely photostable and toxic.

Clay nanoparticles when incorporated into polymer matrices increase reinforcement, leading to stronger plastics, verifiable by a higher glass transition temperature and other mechanical property tests. These nanoparticles are hard, and impart their properties to the polymer (plastic). Nanoparticles have also been attached to textile fibers in order to create smart and functional clothing.

Metal, dielectric, and semiconductor nanoparticles have been formed, as well as hybrid structures (e.g., core–shell nanoparticles). Nanoparticles made of semiconducting material may also be labeled quantum dots if they are small enough (typically sub 10 nm) that quantization of electronic energy levels occurs. Such nanoscale particles are used in biomedical applications as drug carriers or imaging agents.

Semi-solid and soft nanoparticles have been manufactured. A prototype nanoparticle of semi-solid nature is the liposome. Various types of liposome nanoparticles are currently used clinically as delivery systems for anticancer drugs and vaccines.

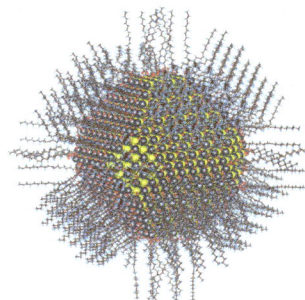

Semiconductor nanoparticle (quantum dot) of lead sulfide with complete passivation by oleic acid, oleyl and hydroxyl (size ~5nm)

Nanoparticles with one half hydrophilic and the other half hydrophobic are termed Janus particles and are particularly effective for stabilizing emulsions. They can self-assemble at water/oil interfaces and act as solid surfactants.

Synthesis

There are several methods for creating nanoparticles, including attrition, pyrolysis and hydrothermal synthesis. In attrition, macro- or micro-scale particles are ground in a ball mill, a planetary ball mill, or other size-reducing mechanism. The resulting particles are air classified to recover nanoparticles. In pyrolysis, a vaporous precursor (liquid or gas) is forced through an orifice at high pressure and burned. The resulting solid (a version of soot) is air classified to recover oxide particles from by-product gases. Traditional pyrolysis often results in aggregates and agglomerates rather than single primary particles. Ultrasonic nozzle spray pyrolysis (USP) on the other hand aids in preventing agglomerates from forming.

A thermal plasma can deliver the energy to vaporize small micrometer-size particles. The thermal plasma temperatures are in the order of 10,000 K, so that solid powder easily evaporates. Nanoparticles are formed upon cooling while exiting the plasma region. The main types of the thermal plasma torches used to produce nanoparticles are dc plasma jet, dc arc plasma, and radio frequency (RF) induction plasmas. In the arc plasma reactors, the energy necessary for evaporation and reaction is provided by an electric arc formed between the anode and the cathode. For example, silica sand can be vaporized with an arc plasma at atmospheric pressure, or thin aluminum wires can be vaporized by exploding wire method. The resulting mixture of plasma gas and silica vapour can be rapidly cooled by quenching with oxygen, thus ensuring the quality of the fumed silica produced.

In RF induction plasma torches, energy coupling to the plasma is accomplished through the electromagnetic field generated by the induction coil. The plasma gas does not come in contact with electrodes, thus eliminating possible sources of contamination and allowing the operation of such plasma torches with a wide range of gases including inert, reducing, oxidizing, and other corrosive atmospheres. The working frequency is typically between 200 kHz and 40 MHz. Laboratory units run at power levels in the order of 30–50 kW, whereas the large-scale industrial units have been tested at power levels up to 1 MW. As the residence time of the injected feed droplets in the plasma is very short, it is important that the droplet sizes are small enough in order to obtain complete evaporation. The RF plasma method has been used to synthesize different nanoparticle materials, for example synthesis of various ceramic nanoparticles such as oxides, carbours/carbides, and nitrides of Ti and Si.

Inert-gas condensation is frequently used to make nanoparticles from metals with low melting points. The metal is vaporized in a vacuum chamber and then supercooled with an inert gas stream. The supercooled metal vapor condenses into nanometer-size particles, which can be entrained in the inert gas stream and deposited on a substrate or studied in situ.

Nanoparticles can also be formed using radiation chemistry. Radiolysis from gamma rays can create strongly active free radicals in solution. This relatively simple technique uses a minimum number of chemicals. These including water, a soluble metallic salt, a radical scavenger (often a secondary alcohol), and a surfactant (organic capping agent). High gamma doses on the order of 10^4 Gray are required. In this process, reducing radicals will drop metallic ions down to the zero-valence state. A scavenger chemical will preferentially interact with oxidizing radicals to prevent the re-oxidation of the metal. Once in the zero-valence state, metal atoms begin to coalesce into particles. A chemical surfactant surrounds the particle during formation and regulates its growth. In sufficient concentrations, the surfactant molecules stay attached to the particle. This prevents it from dissociating or forming clusters with other particles. Formation of nanoparticles using the radiolysis method allows for tailoring of particle size and shape by adjusting precursor concentrations and gamma dose.

Sol-gel

The sol-gel process is a wet-chemical technique (also known as chemical solution deposition) widely used recently in the fields of materials science and ceramic engineering. Such methods are used primarily for the fabrication of materials (typically a metal oxide) starting from a chemical solution (*sol*, short for solution), which acts as the precursor for an integrated network (or *gel*) of either discrete particles or network polymers.

Typical precursors are metal alkoxides and metal chlorides, which undergo hydrolysis and poly-condensation reactions to form either a network "elastic solid" or a colloidal suspension (or dispersion) – a system composed of discrete (often amorphous) submicrometer particles dispersed to various degrees in a host fluid. Formation of a metal oxide involves connecting the metal centers with oxo (M-O-M) or hydroxo (M-OH-M) bridges, therefore generating metal-oxo or met-al-hydroxo polymers in solution. Thus, the sol evolves toward the formation of a gel-like diphasic system containing both a liquid phase and solid phase whose morphologies range from discrete particles to continuous polymer networks.

In the case of the colloid, the volume fraction of particles (or particle density) may be so low that a significant amount of fluid may need to be removed initially for the gel-like properties to be recognized. This can be accomplished in a number of ways. The most simple method is to allow time for sedimentation to occur, and then pour off the remaining liquid. Centrifugation can also be used to accelerate the process of phase separation.

Removal of the remaining liquid (solvent) phase requires a drying process, which is typically causes shrinkage and densification. The rate at which the solvent can be removed is ultimately determined by the distribution of porosity in the gel. The ultimate microstructure of the final component will clearly be strongly influenced by changes implemented during this phase of processing. Afterward, a thermal treatment, or firing process, is often necessary in order to favor further polycondensation and enhance mechanical properties and structural stability via final sintering, densification, and grain growth. One of the distinct advantages of using this methodology as opposed to the more traditional processing techniques is that densification is often achieved at a much lower temperature.

The precursor sol can be either deposited on a substrate to form a film (e.g., by dip-coating or spin-coating), cast into a suitable container with the desired shape (e.g., to obtain a mono-lithic ceramics, glasses, fibers, membranes, aerogels), or used to synthesize powders (e.g., microspheres, nanospheres). The sol-gel approach is a cheap and low-temperature technique that allows for the fine control of the product's chemical composition. Even small quantities of dopants, such as organic dyes and rare earth metals, can be introduced in the sol and end up uniformly dispersed in the final product. It can be used in ceramics processing and manufacturing as an investment casting material, or as a means of producing very thin films of metal oxides for various purposes. Sol-gel derived materials have diverse applications in optics, electronics, energy, space, (bio)sensors, medicine (e.g., controlled drug release) and separation (e.g., chromatography) technology.

Colloids

The term colloid is used primarily to describe a range of mixtures which have solid particles dispersed in a liquid medium. The term applies only if the particles are larger than atomic dimensions but small enough to exhibit Brownian motion. If the particles are large enough, their dynamic behavior in any given period of time in suspension would be governed by forces of gravity and sedimentation. If the particles are small enough, their irregular motion in suspension can be attributed to the collective bombardment of a myriad of thermally agitated molecules in the liquid suspending medium, as described originally by Albert Einstein in his dissertation. Einstein showed evidence that water was made up of discrete molecules by characterizing the observed erratic particle behavior using the theory of Brownian motion, with sedimentation being a possible result. The critical size range (or particle diameter) typically ranges from nanometers (10^{-9} m) to micrometers (10^{-6} m).

Morphology

Scientists have taken to naming their particles after the real-world shapes that they might represent. Nanospheres, nanochains, nanoreefs, nanoboxes and more have appeared in the literature. These morphologies sometimes arise spontaneously as an effect of a templating or directing agent present in the synthesis such as miscellar emulsions or anodized alumina pores, or from the innate crystallographic growth patterns of the materials themselves. Some of these morphologies may serve a purpose, such as long carbon nanotubes used to bridge an electrical junction, or just a scientific curiosity like the stars shown at right.

Nanostars of vanadium(IV) oxide

Amorphous particles usually adopt a spherical shape (due to their microstructural isotropy), whereas the shape of anisotropic microcrystalline whiskers corresponds to their particular crystal habit. At the small end of the size range, nanoparticles are often referred to as clusters. Spheres, rods, fibers, and cups are just a few of the shapes that have been grown. The study of fine particles is called micromeritics.

Characterization

The majority of nanoparticle characterization techniques are light-based, but a non-optical nanoparticle characterization technique called Tunable Resistive Pulse Sensing (TRPS) has been developed that enables the simultaneous measurement of size, concentration and surface charge for a wide variety of nanoparticles. This technique, which applies the Coulter Principle, allows for particle-by-particle quantification of these three nanoparticle characteristics with high resolution.

Functionalization

The surface coating of nanoparticles determines many of their properties, notably stability, solubility, and targeting. A coating that is multivalent or polymeric confers high stability. Functionalized nanomaterial-based catalysts can be used for catalysis of many known organic reactions.

Surface Coating for Biological Applications

For biological applications, the surface coating should be polar to give high aqueous solubility and prevent nanoparticle aggregation. In serum or on the cell surface, highly charged coatings promote non-specific binding, whereas polyethylene glycol linked to terminal hydroxyl or methoxy groups repel non-specific interactions. Nanoparticles can be linked to biological molecules that can act as address tags, to direct the nanoparticles to specific sites within the body, specific organelles within the cell, or to follow specifically the movement of individual protein or RNA molecules in living cells. Common address tags are monoclonal antibodies, aptamers, streptavidin or peptides. These targeting agents should ideally be covalently linked to the nanoparticle and should be present in a controlled number per nanoparticle. Multivalent nanoparticles, bearing multiple targeting groups, can cluster receptors, which can activate cellular signaling pathways, and give stronger anchoring. Monovalent nanoparticles, bearing a single binding site, avoid clustering and so are preferable for tracking the behavior of individual proteins.

Red blood cell coatings can help nanoparticles evade the immune system.

Safety

Nanoparticles present possible dangers, both medically and environmentally. Most of these are due to the high surface to volume ratio, which can make the particles very reactive or catalytic. They are also able to pass through cell membranes in organisms, and their interactions with biological systems are relatively unknown. However, it is unlikely the particles would enter the cell nucleus, Golgi complex, endoplasmic reticulum or other internal cellular components due to the particle size and intercellular agglomeration. A recent study looking at the effects of ZnO nanoparticles on human immune cells has found varying levels of susceptibility to cytotoxicity. There are concerns that pharmaceutical companies, seeking regulatory approval for nano-reformulations of existing medicines, are relying on safety data produced during clinical studies of the earlier, pre-reformulation version of the medicine. This could result in regulatory bodies, such as the FDA, missing new side effects that are specific to the nano-reformulation.

Whether cosmetics and sunscreens containing nanomaterials pose health risks remains largely unknown at this stage. However considerable research has demonstrated that zinc nanoparticles are not absorbed into the bloodstream in vivo.

Concern has also been raised over the health effects of respirable nanoparticles from certain combustion processes. As of 2013 the U.S. Environmental Protection Agency was investigating the safety of the following nanoparticles:

- Carbon Nanotubes: Carbon materials have a wide range of uses, ranging from composites for use in vehicles and sports equipment to integrated circuits for electronic components. The interactions between nanomaterials such as carbon nanotubes and natural organic matter strongly influence both their aggregation and deposition, which strongly affects their transport, transformation, and exposure in aquatic environments. In past research, carbon nanotubes exhibited some toxicological impacts that will be evaluated in various environmental settings in current EPA chemical safety research. EPA research will provide data, models, test methods, and best practices to discover the acute health effects of carbon nanotubes and identify methods to predict them.

- Cerium oxide: Nanoscale cerium oxide is used in electronics, biomedical supplies, energy, and fuel additives. Many applications of engineered cerium oxide nanoparticles naturally disperse themselves into the environment, which increases the risk of exposure. There is ongoing exposure to new diesel emissions using fuel additives containing CeO_2 nanoparticles, and the environmental and public health impacts of this new technology are unknown. EPA's chemical safety research is assessing the environmental, ecological, and health implications of nanotechnology-enabled diesel fuel additives.

- Titanium dioxide: Nano titanium dioxide is currently used in many products. Depending on the type of particle, it may be found in sunscreens, cosmetics, and paints and coatings. It is also being investigated for use in removing contaminants from drinking water.

- Nano Silver: Nano silver is being incorporated into textiles, clothing, food packaging, and other materials to eliminate bacteria. EPA and the U.S. Consumer Product Safety Commission are studying certain products to see whether they transfer nano-size silver particles in real-world scenarios. EPA is researching this topic to better understand how much nano-silver children come in contact with in their environments.

- Iron: While nano-scale iron is being investigated for many uses, including "smart fluids" for uses such as optics polishing and as a better-absorbed iron nutrient supplement, one of its more prominent current uses is to remove contamination from groundwater. This use, supported by EPA research, is being piloted at a number of sites across the country.

Laser Applications

The use of nanoparticles in laser dye-doped poly(methyl methacrylate) (PMMA) laser gain media was demonstrated in 2003 and it has been shown to improve conversion efficiencies and to decrease laser beam divergence. Researchers attribute the reduction in beam divergence to improved dn/dT characteristics of the organic-inorganic dye-doped nanocomposite. The optimum composition reported by these researchers is 30% w/w of SiO_2 (~ 12 nm) in dye-doped PMMA.

Medicinal Applications

- Liposome
- Dendrimer

- Iron oxide nanoparticles

- Nanomedicine

- Polymer-drug conjugate

- Polymeric nanoparticle

Nanoparticle–biomolecule Conjugate

A nanoparticle–biomolecule conjugate is a nanoparticle with biomolecules attached to its surface. Nanoparticles are minuscule particles, typically measured in nanometers (nm), that are used in nanobiotechnology to explore the functions of biomolecules. Properties of the ultrafine particles are characterized by the components on their surfaces more so than larger structures, such as cells, due to large surface area-to-volume ratios. Large surface area-to-volume-ratios of nanoparticles optimize the potential for interactions with biomolecules.

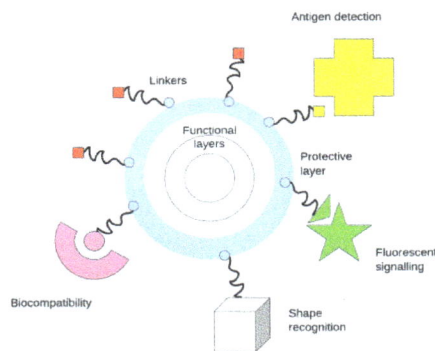

Attachments on nanoparticles make them more biocompatible.

Characterization

Major characteristics of nanoparticles include volume, structure, and visual properties that make them valuable in nanobiotechnology. Depending on specific properties of size, structure, and luminescence, nanoparticles can be used for different applications. Imaging techniques are used to identify such properties and give more information about the tested sample. Techniques used to characterize nanoparticles are also useful in studying how nanoparticles interact with biomolecules, such as amino acids or DNA, and include:

- Magnetic resonance imaging (MRI), denoted by the solubility of the nanoparticles in water and fluorescent. MRI can be applied in the medical field to visualize structures.

- Atomic force microscopy (AFM) gives the researcher or scientist a topographic view of the sample on a substrate.

- Transmission electron microscopy (TEM) gives a top view, but with a different technique then that of atomic force microscopy.

- Raman spectroscopy or surface enhanced raman spectroscopy (SERS) gives information about wavelengths and energy in the sample.

- Ultra-violet spectroscopy (UV-Vis) measures the wavelengths where light is absorbed.

- X-Ray diffraction (XRD) generally gives an idea of the chemical composition of the sample.

Chemistry

Physical

Nanomolecules can be created from virtually any element, but the majority produced in today's industry use carbon as the basis upon which the molecules are built around. Carbon can bond with nearly any element, allowing many possibilities when it comes to creating a specific molecule. Scientists can create thousands upon thousands of individual nanomolecules from a simple carbon basis. Some of the most famous nanomolecules currently in existence are solely carbon; these include carbon nanotubes and buckminsterfullerenes. In contrast with nanomolecules, the chemical components of nanoparticles usually consist of metals, such as iron, gold, silver, and platinum.

Interactions between nanoparticles and molecules change depending on the nanoparticle's core. Nanoparticle properties depend not only on the composition of the core material, but also on varying thicknesses of material used. Magnetic properties are particularly useful in molecule manipulation, and thus metals are often used as core material. Metals contain inherent magnetic properties that allow for manipulation of molecular assembly. As nanoparticles interact with molecules via ligand properties, molecular assembly can be controlled by external magnetic fields interacting with magnetic properties in the nanoparticles. Significant problems with producing nanoparticles initially arise once these nanoparticles are generated in solution. Without the use of a stabilizing agent, nanoparticles tend to stick together once the stirring is stopped. In order to counteract this, a certain collidial stabilizer is generally added. These stabilizers bind to the nanoparticles in a way that prevents other particles from bonding with them. Some effective stabilizers found so far include citrate, cellulose, and sodium borohydride.

Application Chemistry

Nanoparticles are desirable in today's industry for their high surface area-to-volume ratio in comparison with larger particles of the same elements. Because chemical reactions occur at a rate directly proportional to the available surface area of reactant compounds, nanoparticles can generate reactions at a much faster rate than larger particles of equal mass. Nanoparticles therefore are among the most efficient means of producing reactions and are inherently valuable in the chemical industry. The same property makes them valuable in interactions with molecules.

Applications with Biomolecules and Biological Processes

Nanoparticles have the potential to greatly influence biological processes. The potency of a nanoparticle increases as its surface area-to-volume-ratio does. Attachments of ligands to the surface of nanoparticles allow them to interact with biomolecules.

Identification of Biomolecules

Nanoparticles are valuable tools in identification of biomolecules, through the use of bio-tagging or labeling. Attachments of ligands or molecular coatings to the surface of a nanoparticle facilitate nanoparticle-molecule interaction, and make them biocompatible. Conjugation can be achieved through intermolecular attractions between the nanoparticle and biomolecule such as:

- Covalent bonding

- Chemisorption

- Noncovalent interactions

To enhance visualization, nanoparticles can also be made to fluoresce by controlling the size and shape of a nanoparticle probe. Fluorescence increases luminescence by increasing the range of wavelengths the emitted light can reach, allowing for biomarkers with a variety of colors. This technique is used to track the efficacy of protein transfer both in vivo and in vitro in terms of genetic alternations.

Biological Process Control

Biological processes can be controlled through transcription regulation, gene regulation, and enzyme inhibition processes that can be regulated using nanoparticles. Nanoparticles can play a part in gene regulation through ionic bonding between positively charged cationic ligands on the surfaces of nanoparticles and negatively charged anionic nucleic acids present in DNA. In an experiment, a nanoparticle-DNA complex inhibited transcription by T7 RNA polymerase, signifying strong bonding in the complex. A high affinity of the nanoparticle-DNA complex indicates strong bonding and a favorable use of nanoparticles. Attaching ionic ligands to nanoparticles allows control over enzyme activity. An example of enzyme inhibition is given by binding of a-chymotrypsin (ChT), an enzyme with a largely cationic active site. When a-chymotrypsin is incubated with anionic (negatively charged) nanoparticles, ChT activity is inhibited as anionic nanoparticles bind to the active site. Enzyme activity can be restored by the addition of cationic surfactants. Alkyl surfactants form a bilayer around ChT, whereas thiol and alcohol surfactants alter the surface of ChT such that interactions with nanoparticles are interrupted. Though formation of a protein-nanoparticle complex can inhibit enzyme activity, studies show that it can also stabilize protein structure, and significantly protect the protein from denaturization. Attachments of ligands to segments of nanoparticles selected for functionalization of metallic properties can be used to generate a magnetic nanowire, which generates a magnetic field that allows for the manipulation of cellular assemblies.

Genetic Alteration

Nanoparticles can also be used in conjunction with DNA to perform genetic alterations. These are frequently monitored through the use of fluorescent materials, allowing scientists to judge if these tagged proteins have successfully been transmitted—for example Green fluorescent protein, or GFP. Nanoparticles are significantly less cytotoxic than currently used organic methods, providing a more efficient method of monitoring genetic alternations. They also do not degrade or bleach with time, as organic dyes do. Suspensions of nanoparticles with the same size and shapes (mono-dispersed) with functional groups attached to their surfaces can also electrostatically bind to DNA, protecting them from several types of degradation. Because the fluorescence of these nanoparticles does not degrade, cellular localization can be tracked without the use of additional tagging, with GFPs or other methods. The 'unpacking' of the DNA can be detected in live cells using luminescence resonance energy transfer (LRET) technology.

Medical Implications

Small molecules *in vivo* have a short retention time, but the use of larger nanoparticles does not. These nanoparticles can be used to avoid immune response, which aids in the treatment of chronic diseases. It has been investigated as a potential cancer therapy and also has the potential to affect the understanding of genetic disorders. Nanoparticles also have the potential to aid in site-specific drug delivery by improving the amount of unmodified drug that is circulated within the system, which also decreases the necessary dosage frequency. The targeted nature of nanoparticles also means that non-targeted organs are much less likely to experience side effects of drugs intended for other areas.

Another example is the delivery of nucleic acids to alter key genes associated with cancer progression. For example, Conde et al have shown that by twisting RNA strands into a triple helix and embedding them in a biocompatible gel, they can not only deliver the strands efficiently but also use them to shrink aggressive tumors in mice. Using this technique, the researchers dramatically improved cancer survival rates by simultaneously turning on a tumor-suppressing microRNA and de-activating one that causes cancer. They believe their approach could also be used for delivering other types of RNA, as well as DNA and other therapeutic molecules.

Studying Cell Interactions

Cellular interactions occur at a microscopic level and cannot be easily observed even with the advanced microscopes available today. Due to difficulties observing reactions at the molecular level, indirect methods are used which greatly limits the scope of the understanding that can be gained by studying these processes essential to life. Advances in the material industry has evolved a new field known as nanobiotechnology, that uses nanoparticles to study interactions at the biomolecular level.

One area of research featuring nanobiotechnology is the extracellular matrices of cells (ECM). The ECM is primarily composed of interwoven fibers of collagen and elastin that have diameters ranging from 10-300 nm. In addition to holding the cell in place, the ECM has a variety of other functions including providing a point of attachment for the ECM of other cells and transmembrane receptors that are essential for life. Until recently it has been nearly impossible to study the physical forces that help cells maintain their functionality, but nanobiotechnology has given us the ability to learn more about these interactions. Using the unique properties of nanoparticles, it is possible to control how the nanoparticles adhere to certain patterns present in the ECM, and as a result can understand how changes in the ECM's shape can affect cell functionality.

Using nanobiotechnology to study the ECM allows scientists to investigate the binding interactions that occur between the ECM and its supporting environment. Investigators were able to study these interactions by utilizing tools such as optical tweezers, which have the ability to trap nano-scale objects with focused light. The tweezers can affect the binding of a substrate to the ECM by attempting to draw the substrate away from it. The light emitted from the tweezers was used to restrain ECM-coated microbeads, and the changes in the force exerted by the ECM onto the substrate were studied by modulating the effect of the optical tweezers. Experiments showed that the force exerted by the ECM on the substrate positively correlated with the force of the tweezers, which led to the subsequent discovery that the ECM and the transmembrane proteins are able to sense external forces, and can adapt to overcome these forces.

Nanotechnology Crossing the Blood-brain Barrier

The blood-brain barrier (BBB) is composed of a system of capillaries that has an especially dense lining of endothelial cells which protects the central nervous system (CNS) against the diffusion of foreign substances into the cerebrospinal fluid. These harmful objects include microscopic bacteria, large hydrophobic molecules, certain hormones and neurotransmitters, and low-lipid-soluble molecules. The BBB prevents these harmful particles from entering the brain via tight junctions between endothelial cells and metabolic barriers. The thoroughness with which the BBB does its job makes it difficult to treat diseases of the brain such as cancer, Alzheimer's, and autism, because it is very difficult to transport drugs across the BBB. Currently, in order to deliver therapeutic molecules into the brain, doctors must use highly invasive techniques such as drilling directly into the brain, or sabotaging the integrity of the BBB through biochemical means. Due to their small size and large surface area, nanoparticles offer a promising solution for neurotherapeutics.

Nanotechnology is helpful in delivering drugs and other molecules across the blood-brain barrier (BBB). Nanoparticles allow drugs, or other foreign molecules, to efficiently cross the BBB by camouflaging themselves and tricking the brain into providing them with the ability to cross the BBB in a process called the Trojan Horse Method. Using nanotechnology is advantageous because only the engineered complex is necessary whereas in ordinary applications the active compound must carry out the reaction. This allows for maximum efficacy of the active drug. Also, the use of nanoparticles results in the attraction of proteins to the surfaces of cells, giving cell membranes a biological identity. They also use endogenous active transport where Transferrin, an iron binding protein, is linked to rod-shaped semiconductor nanocrystals, in order to move across the BBB into the brain. This discovery is a promising development towards designing an efficient nanoparticle-based drug delivery system.

References

- MacNaught, Alan D.; Wilkinson, Andrew R., eds. (1997). Compendium of Chemical Terminology: IUPAC Recommendations (2nd ed.). Blackwell Science. ISBN 0865426848.

- Reiss, Gunter; Hutten, Andreas (2010). "Magnetic Nanoparticles". In Sattler, Klaus D. Handbook of Nanophysics: Nanoparticles and Quantum Dots. CRC Press. pp. 2–1. ISBN 9781420075458.

- Corriu, Robert & Anh, Nguyên Trong (2009). Molecular Chemistry of Sol-Gel Derived Nanomaterials. John Wiley and Sons. ISBN 0-470-72117-0.

- Rotello, Vincent (2004). Nanoparticles Building Blocks for Nanotechnology. New York: Springer Science + Business Media Inc. ISBN 0306482878.

- "Toxic Nanoparticles Might be Entering Human Food Supply, MU Study Finds". University of Missouri. 22 August 2013. Retrieved 23 August 2013.

- Benson, H., Sarveiya, V., Risk, S. and Roberts, M. S. (2005) . "Influence of anatomical site and topical for-

mulation on skin penetration of sunscreens. Therapeutics and Clinical Risk Management, 1 3: 209–218."
Retrieved 2012-04-01.

- Howard, V. (2009). "Statement of Evidence: Particulate Emissions and Health (An Bord Plenala, on Proposed Ringaskiddy Waste-to-Energy Facility)." Retrieved 2011-04-26.

- Salata, OV (30 April 2004). "Applications of Nanoparticles in Biology and Medicine". Journal of Nanobiotechnology. 2 (1). doi:10.1186/1477-3155-2-3. PMC 419715. PMID 15119954. Retrieved 29 April 2011.

- Sniadecki, Nathan; Ravi Desai; Sami Ruiz; Christopher Chen (1 January 2006). "Nanotechnology for Cell–Substrate Interactions". Annals of Biomedical Engineering. 34 (1): 59–74. doi:10.1007/s10439-005-9006-3. Retrieved 2011-04-29.

- Raghnaill, Michelee; Meredith Brown; Dong Ye (April 2011). "Internal benchmarking of a human blood-brain barrier cell model for screening of nanoparticle uptake and transcytosis". European Journal of Pharmaceutics and Biopharmaceutics. 77 (3): 360–367. doi:10.1016/j.ejpb.2010.10.024. Retrieved April 29, 2011.

- Greulich, C.; Diendorf, J.; Simon, T.; Eggeler, G.; Epple, M.; Köller, M. (2011). "Uptake and intracellular distribution of silver nanoparticles in human mesenchymal stem cells". Acta Biomaterialia. 7 (1): 347–354. doi:10.1016/j.actbio.2010.08.003. ISSN 1742-7061.

3

Various Types of Nanocomposites

Nanocomposites can be classified under metal matrix-based nanocomposites, polymer matrix nanocomposites and ceramic matrix nanocomposites. This chapter provides a detailed account of the varying chemical, structural, thermal and electrical conductivities and magnetic susceptibilities of the various types of nanocomposites. The chapter also deals with topics like carbon nanotube, silver nanoparticle, hybrid material, nanowire and nanocellulose.

Polymer Nanocomposite

Polymer nanocomposites (PNC) consist of a polymer or copolymer having nanoparticles or nanofillers dispersed in the polymer matrix. These may be of different shape (e.g., platelets, fibers, spheroids), but at least one dimension must be in the range of 1–50 nm. These PNC's belong to the category of multi-phase systems (MPS, viz. blends, composites, and foams) that consume nearly 95% of plastics production. These systems require controlled mixing/compounding, stabilization of the achieved dispersion, orientation of the dispersed phase, and the compounding strategies for all MPS, including PNC, are similar.

Polymer nanoscience is the study and application of nanoscience to polymer-nanoparticle matrices, where nanoparticles are those with at least one dimension of less than 100 nm.

The transition from micro- to nano-particles lead to change in its physical as well as chemical properties. Two of the major factors in this are the increase in the ratio of the surface area to volume, and the size of the particle. The increase in surface area-to-volume ratio, which increases as the particles get smaller, leads to an increasing dominance of the behavior of atoms on the surface area of particle over that of those interior of the particle. This affects the properties of the particles when they are reacting with other particles. Because of the higher surface area of the nano-particles, the interaction with the other particles within the mixture is more and this increases the strength, heat resistance, etc. and many factors do change for the mixture.

An example of a nanopolymer is silicon nanospheres which show quite different characteristics; their size is 40–100 nm and they are much harder than silicon, their hardness being between that of sapphire and diamond.

Bio-hybrid Polymer Nanofibers

Many technical applications of biological objects like proteins, viruses or bacteria such as chromatography, optical information technology, sensorics, catalysis and drug delivery require their immobilization. Carbon nanotubes, gold particles and synthetic polymers are used for this purpose. This immobilization has been achieved predominantly by adsorption or by chemical binding and to a lesser extent by incorporating these objects as guests in host matrices. In the guest host systems, an ideal method for the immobilization of biological objects and their integration into hierarchical architectures should be structured on a nanoscale to facilitate the interactions of biological nano-objects with their environment. Due to the large number of natural or synthetic polymers available and the advanced techniques developed to process such systems to nanofibres, rods, tubes etc. make polymers a good platform for the immobilization of biological objects.

Bio-Hybrid Nanofibres by Electrospinning

Polymer fibers are, in general, produced on a technical scale by extrusion, i.e., a polymer melt or a polymer solution is pumped through cylindrical dies and spun/drawn by a take-up device. The resulting fibers have diameters typically on the 10-μm scale or above. To come down in diameter into the range of several hundreds of nanometers or even down to a few nanometers, Electrospinning is today still the leading polymer processing technique available. A strong electric field of the order of 103 V/cm is applied to the polymer solution droplets emerging from a cylindrical die. The electric charges, which are accumulated on the surface of the droplet, cause droplet deformation along the field direction, even though the surface tension counteracts droplet evolution. In super-critical electric fields, the field strength overbears the surface tension and a fluid jet emanates from the droplet tip. The jet is accelerated towards the counter electrode. During this transport phase, the jet is subjected to strong electrically driven circular bending motions that cause a strong elongation and thinning of the jet, a solvent evaporation until, finally, the solid nanofibre is deposited on the counter electrode.

Bio-Hybrid Polymer Nanotubes by Wetting

Electro spinning, co-electrospinning, and the template methods based on nanofibres yield nano-objects which are, in principle, infinitively long. For a broad range of applications including catalysis, tissue engineering, and surface modification of implants this infinite length is an advantage. But in some applications like inhalation therapy or systemic drug delivery, a well-defined length is required. The template method to be described in the following has the advantage such that it allows the preparation of nanotubes and nanorods with very high precision. The method is based on the use of well defined porous templates, such as porous aluminum or silicon.

The basic concept of this method is to exploit wetting processes. A polymer melt or solution is brought into contact with the pores located in materials characterized by high energy surfaces such as aluminum or silicon. Wetting sets in and covers the walls of the pores with a thin film with a thickness of the order of a few tens of nanometers.

Gravity does not play a role, as it is obvious from the fact that wetting takes place independent of the orientation of the pores relative to the direction of gravity. The exact process is still not understood theoretically in detail but its known from experiments that low molar mass systems tend to fill the pores completely, whereas polymers of sufficient chain length just cover the walls. This process happens typically within a minute for temperatures about 50 K above the melting temperature or glass transition temperature, even for highly viscous polymers, such as, for instance, polytetrafluoroethylene, and this holds even for pores with an aspect ratio as large as 10,000. The complete filling, on the other hand, takes days. To obtain nanotubes, the polymer/template system is cooled down to room temperature or the solvent is evaporated, yielding pores covered with solid layers. The resulting tubes can be removed by mechanical forces for tubes up to 10 μm in length, i.e., by just drawing them out from the pores or by selectively dissolving the template. The diameter of the nanotubes, the distribution of the diameter, the homogeneity along the tubes, and the lengths can be controlled.

Applications

The nanofibres, hollow nanofibres, core–shell nanofibres, and nanorods or nanotubes produced have a great potential for a broad range of applications including homogeneous and heterogeneous catalysis, sensorics, filter applications, and optoelectronics. Here we will just consider a limited set of applications related to life science.

Tissue Engineering

This is mainly concerned with the replacement of tissues which have been destroyed by sickness or accidents or other artificial means. The examples are skin, bone, cartilage, blood vessels and may be even organs. This technique involves providing a scaffold on which cells are added and the scaffold should provide favorable conditions for the growth of the same. Nanofibres have been found to provide very good conditions for the growth of such cells, one of the reasons being that fibrillar structures can be found on many tissues which allow the cells to attach strongly to the fibers and grow along them as shown.

Nanoparticles such as graphene, carbon nanotubes, molybdenum disulfide and tungsten disulfide are being used as reinforcing agents to fabricate mechanically strong biodegradable polymeric nanocomposites for bone tissue engineering applications. The addition of these nanoparticles in the polymer matrix at low concentrations (~0.2 weight %) leads to significant improvements in the compressive and flexural mechanical properties of polymeric nanocomposites. Potentially, these nanocomposites may be used as a novel, mechanically strong, light weight composite as bone implants. The results suggest that mechanical reinforcement is dependent on the nanostructure morphology, defects, dispersion of nanomaterials in the polymer matrix, and the cross-linking density of the polymer. In general, two-dimensional nanostructures can reinforce the polymer better than one-dimensional nanostructures, and inorganic nanomaterials are better reinforcing agents than carbon based nanomaterials.

Delivery from Compartmented Nanotubes

Nano tubes are also used for carrying drugs in general therapy and in tumor therapy in particular. The role of them is to protect the drugs from destruction in blood stream, to control the delivery with a well-defined release kinetics, and in ideal cases, to provide vector-targeting properties or release mechanism by external or internal stimuli.

Rod or tube-like, rather than nearly spherical, nanocarriers may offer additional advantages in terms of drug delivery systems. Such drug carrier particles possess additional choice of the axial ratio, the curvature, and the "all-sweeping" hydrodynamic-related rotation, and they can be modified chemically at the inner surface, the outer surface, and at the end planes in a very selective way. Nanotubes prepared with a responsive polymer attached to the tube opening allow the control of access to and release from the tube. Furthermore, nanotubes can also be prepared showing a gradient in its chemical composition along the length of the tube.

Compartmented drug release systems were prepared based on nanotubes or nanofibres. Nanotubes and nanofibres, for instance, which contained fluorescent albumin with dog-fluorescein isothiocyanate were prepared as a model drug, as well as super paramagnetic nanoparticles composed of iron oxide or nickel ferrite. The presence of the magnetic nanoparticles allowed, first of all, the guiding of the nanotubes to specific locations in the body by external magnetic fields. Super paramagnetic particles are known to display strong interactions with external magnetic fields leading to large saturation magnetizations. In addition, by using periodically varying magnetic fields, the nanoparticles were heated up to provide, thus, a trigger for drug release. The presence of the model drug was established by fluorescence spectroscopy and the same holds for the analysis of the model drug released from the nanotubes.

Immobilization of Proteins

Core shell fibers of nano particles with fluid cores and solid shells can be used to entrap biological objects such as proteins, viruses or bacteria in conditions which do not affect their functions. This effect can be used among others for biosensor applications. For example, Green Fluorescent Protein is immobilized in nanostructured fibres providing large surface areas and short distances for the analyte to approach the sensor protein.

With respect to using such fibers for sensor applications fluorescence of the core shell fibers was found to decay rapidly as the fibers were immersed into a solution containing urea: urea permeates through the wall into the core where it causes denaturation of the GFP. This simple experiment reveals that core–shell fibers are promising objects for preparing biosensors based on biological objects.

Polymer nanostructured fibers, core–shell fibers, hollow fibers, and nanorods and nanotubes provide a platform for a broad range of applications both in material science as well as in life science. Biological objects of different complexity and synthetic objects carrying specific functions can be incorporated into such nanostructured polymer systems while keeping their specific functions vital. Biosensors, tissue engineering, drug delivery, or enzymatic catalysis is just a few of the possible examples. The incorporation of viruses and bacteria all the way up to microorganism should not really pose a problem and the applications coming from such biohybrid systems should be tremendous.

Size and Pressure Effects on Nanopolymers

The size- and pressure- dependent glass transition temperatures of free-standing films or supported films having weak interactions with **substrates** decreases with decreasing of pressure and size. However, the glass transition temperature of supported films having strong interaction with substrates increases of pressure and the decrease of size. Different models like two layer model, three layer model, T_g (D, 0) 1/D and some more models relating specific heat, density and thermal expansion are used to obtain the experimental results on nanopolymers and even some observations like freezing of films due to memory effects in the visco-elastic eigenmodels of the films, and finite effects of the small molecule glass are observed. To describe T_g (D, 0) function of polymers more generally, a simple and unified model recently is provided based on the size-dependent melting temperature of crystals and Lindemann's criterion

$$T_g \, (D, 0) \, / \, T_g \, (\infty, 0) \; \sigma_g^2 \, (\infty, 0) \, / \, \sigma_g^2 \, (D, 0)$$

where σ_g is the root of mean squared displacement of surface and interior molecules of glasses at T_g (D, 0), $\alpha = \sigma_s^2$ (D, 0) / σ_v^2 (D, 0) with subscripts s and v denoting surface and volume, respectively. For a nanoparticle, D has a usual meaning of diameter, for a nanowire, D is taken as its diameter, and for a thin film, D denotes its thickness. D_o denotes a critical diameter at which all molecules of a low-dimensional glass are located on its surface.

Conclusion

The devices that use the properties of low-dimensional objects such as nanoparticles are promising due to the possibility of tailoring a number of electrophysical, optical and magnetic properties changing the size of nanoparticles, which can be controlled during the synthesis. In the case of polymer nanocomposites we can use the properties of disordered systems.

Here recent developments in the field of polymer nano-composites and some of their applications have been reviewed. Though there is much use in this field, there are many limitations also. For example, in the release of drugs using nanofibres, cannot be controlled independently and a burst release is usually the case, whereas a more linear release is required. Let us now consider future aspects in this field.

There is a possibility of building ordered arrays of nanoparticles in the polymer matrix. A number of possibilities also exist to manufacture the nanocomposite circuit boards. An even more attractive method exists to use polymer nanocomposites for neural networks applications. Another promising area of development is optoelectronics and optical computing. The single domain nature and super paramagnetic behavior of nanoparticles containing ferromagnetic metals could be possibly used for magneto-optical storage media manufacturing.

Carbon Nanotube

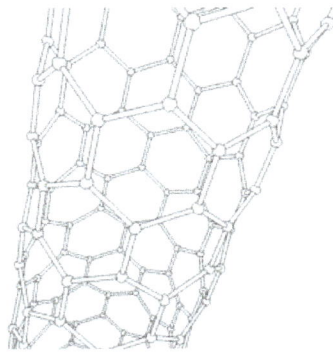

Rotating single-walled zigzag carbon nanotube

Carbon nanotubes (CNTs) are allotropes of carbon with a cylindrical nanostructure. Nanotubes have been constructed with length-to-diameter ratio of up to 132,000,000:1, significantly larger than for any other material. These cylindrical carbon molecules have unusual properties, which are valuable for nanotechnology, electronics, optics and other fields of materials science and technology. In particular, owing to their extraordinary thermal conductivity and mechanical and electrical properties, carbon nanotubes find applications as additives to various structural materials.

For instance, nanotubes form a tiny portion of the material(s) in some (primarily carbon fiber) baseball bats, golf clubs, car parts or damascus steel.

Nanotubes are members of the fullerene structural family. Their name is derived from their long, hollow structure with the walls formed by one-atom-thick sheets of carbon, called graphene. These sheets are rolled at specific and discrete ("chiral") angles, and the combination of the rolling angle and radius decides the nanotube properties; for example, whether the individual nanotube shell is a metal or semiconductor. Nanotubes are categorized as single-walled nanotubes (SWNTs) and multi-walled nanotubes (MWNTs). Individual nanotubes naturally align themselves into "ropes" held together by van der Waals forces, more specifically, pi-stacking.

Applied quantum chemistry, specifically, orbital hybridization best describes chemical bonding in nanotubes. The chemical bonding of nanotubes is composed entirely of sp^2 bonds, similar to those of graphite. These bonds, which are stronger than the sp^3 bonds found in alkanes and diamond, provide nanotubes with their unique strength.

Types of Carbon Nanotubes and Related Structures

Terminology

There is no consensus on some terms describing carbon nanotubes in scientific literature: both "-wall" and "-walled" are being used in combination with "single", "double", "triple" or "multi", and the letter C is often omitted in the abbreviation; for example, multi-walled carbon nanotube (MWNT).

Single-walled

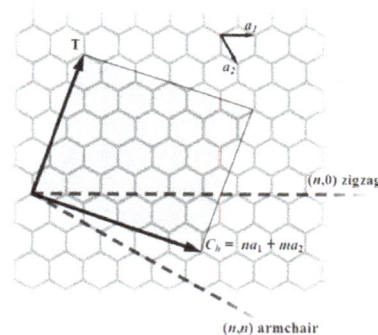

The (n,m) nanotube naming scheme can be thought of as a vector ($\mathbf{C}_h$) in an infinite graphene sheet that describes how

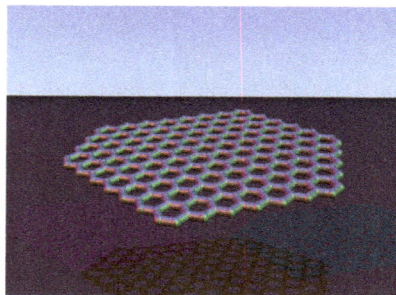

Graphene nanoribbon to "roll up" the graphene sheet to make the nanotube. $\mathbf{T}$ denotes the tube axis, and $\mathbf{a}_1$ and $\mathbf{a}_2$ are the unit vectors of graphene in real space.

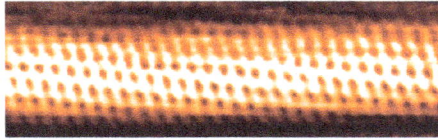

A scanning tunneling microscopy image of single-walled carbon nanotube

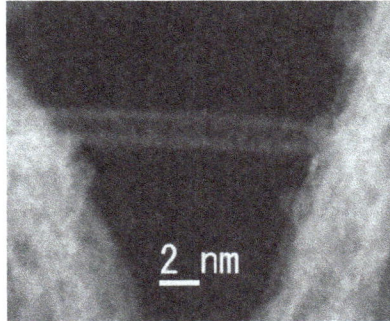

A transmission electron microscopy image of a single-walled carbon nanotube

Most single-walled nanotubes (SWNTs) have a diameter of close to 1 nanometer, and can be many millions of times longer. The structure of a SWNT can be conceptualized by wrapping a one-atom-thick layer of graphite called graphene into a seamless cylinder. The way the graphene sheet is wrapped is represented by a pair of indices (n,m). The integers n and m denote the number of unit vectors along two directions in the honeycomb crystal lattice of graphene. If $m = 0$, the nanotubes are called zigzag nanotubes, and if $n = m$, the nanotubes are called armchair nanotubes. Otherwise, they are called chiral. The diameter of an ideal nanotube can be calculated from its (n,m) indices as follows

$$d = \frac{a}{\pi}\sqrt{(n^2 + nm + m^2)} = 78.3\sqrt{((n+m)^2 - nm)}\text{pm},$$

where a = 0.246 nm.

SWNTs are an important variety of carbon nanotube because most of their properties change significantly with the (n,m) values, and this dependence is non-monotonic. In par-ticular, their band gap can vary from zero to about 2 eV and their electrical conductivity can show metallic or semiconducting behavior. Single-walled nanotubes are likely candidates for miniatur-izing electronics. The most basic building block of these systems is the electric wire, and SWNTs with diameters of an order of a nanometer can be excellent conductors. One useful application of SWNTs is in the development of the first intermolecular field-effect transistors (FET). The first in-termolecular logic gate using SWCNT FETs was made in 2001. A logic gate requires both a p-FET and an n-FET. Because SWNTs are p-FETs when exposed to oxygen and n-FETs otherwise, it is possible to expose half of an SWNT to oxygen and protect the other half from it. The resulting SWNT acts as a *not* logic gate with both p and n-type FETs in the same molecule.

Prices for single-walled nanotubes declined from around $1500 per gram as of 2000 to retail prices of around $50 per gram of as-produced 40–60% by weight SWNTs as of March 2010. As of 2016 the retail price of as-produced 75% by weight SWNTs were $2 per gram, cheap enough for widespread use. SWNTs are forecast to make a large impact in electronics applications by 2020 according to the *The Global Market for Carbon Nanotubes* report.

Multi-walled

Multi-walled nanotubes (MWNTs) consist of multiple rolled layers (concentric tubes) of graphene. There are two models that can be used to describe the structures of multi-walled nanotubes. In the *Russian Doll* model, sheets of graphite are arranged in concentric cylinders, e.g., a (0,8) single-walled nanotube (SWNT) within a larger (0,17) single-walled nanotube. In the *Parchment* model, a single sheet of graphite is rolled in around itself, resembling a scroll of parchment or a rolled newspaper. The interlayer distance in multi-walled nanotubes is close to the distance between graphene layers in graphite, approximately 3.4 Å. The Russian Doll structure is observed more commonly. Its individual shells can be described as SWNTs, which can be metallic or semiconducting. Because of statistical probability and restrictions on the relative diameters of the individual tubes, one of the shells, and thus the whole MWNT, is usually a zero-gap metal.

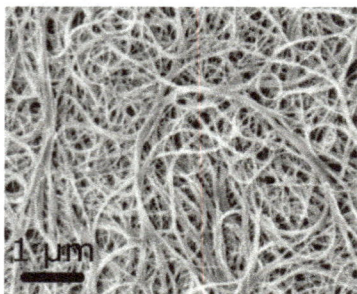

A scanning electron microscopy image of carbon nanotubes bundles

Double-walled carbon nanotubes (DWNTs) form a special class of nanotubes because their morphology and properties are similar to those of SWNTs but they are more resistant to chemicals. This is especially important when it is necessary to graft chemical functions to the surface of the nanotubes (functionalization) to add properties to the CNT. Covalent functionalization of SWNTs will break some C=C double bonds, leaving "holes" in the structure on the nanotube, and thus modifying both its mechanical and electrical properties. In the case of DWNTs, only the outer wall is modified. DWNT synthesis on the gram-scale was first proposed in 2003 by the CCVD technique, from the selective reduction of oxide solutions in methane and hydrogen.

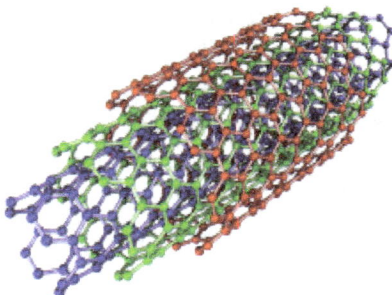

Triple-walled armchair carbon nanotube

The telescopic motion ability of inner shells and their unique mechanical properties will permit the use of multi-walled nanotubes as main movable arms in coming nanomechanical devices. Retraction force that occurs to telescopic motion caused by the Lennard-Jones interaction between shells and its value is about 1.5 nN.

Torus

In theory, a nanotorus is a carbon nanotube bent into a torus (doughnut shape). Nanotori are predicted to have many unique properties, such as magnetic moments 1000 times larger than previously expected for certain specific radii. Properties such as magnetic moment, thermal stability, etc. vary widely depending on radius of the torus and radius of the tube.

Nanobud

Carbon nanobuds are a newly created material combining two previously discovered allotropes of carbon: carbon nanotubes and fullerenes. In this new material, fullerene-like "buds" are covalently bonded to the outer sidewalls of the underlying carbon nanotube. This hybrid material has useful properties of both fullerenes and carbon nanotubes. In particular, they have been found to be exceptionally good field emitters. In composite materials, the attached fullerene molecules may function as molecular anchors preventing slipping of the nanotubes, thus improving the composite's mechanical properties.

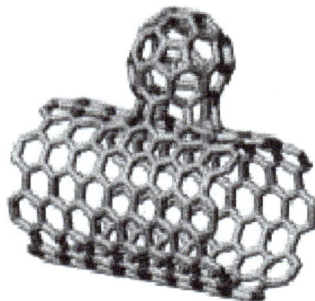

A stable nanobud structure

Three-dimensional Carbon Nanotube Architectures

Recently, several studies have highlighted the prospect of using carbon nanotubes as building blocks to fabricate three-dimensional macroscopic (>100 nm in all three dimensions) all-carbon devices. Lalwani et al. have reported a novel radical initiated thermal crosslinking method to fabricate macroscopic, free-standing, porous, all-carbon scaffolds using single- and multi-walled carbon nanotubes as building blocks. These scaffolds possess macro-, micro-, and nano- structured pores and the porosity can be tailored for specific applications. These 3D all-carbon scaffolds/architectures may be used for the fabrication of the next generation of energy storage, supercapacitors, field emission transistors, high-performance catalysis, photovoltaics, and biomedical devices and implants.

3D carbon scaffolds

Graphenated Carbon Nanotubes (g-CNTs)

SEM series of graphenated CNTs with varying foliate density

Graphenated CNTs are a relatively new hybrid that combines graphitic foliates grown along the sidewalls of multiwalled or bamboo style CNTs. Yu *et al.* reported on "chemically bonded graphene leaves" growing along the sidewalls of CNTs. Stoner *et al.* described these structures as "graphenated CNTs" and reported in their use for enhanced supercapacitor performance. Hsu *et al.* further reported on similar structures formed on carbon fiber paper, also for use in supercapacitor applications. Pham *et al.* also reported a similar structure, namely "graphene-carbon nanotube hybrids", grown directly onto carbon fiber paper to form an integrated, binder free, high surface area conductive catalyst support for Proton Exchange Membrane Fuel Cells electrode applications with enhanced performance and durability. The foliate density can vary as a function of deposition conditions (e.g. temperature and time) with their structure ranging from few layers of graphene (< 10) to thicker, more graphite-like.

The fundamental advantage of an integrated graphene-CNT structure is the high surface area three-dimensional framework of the CNTs coupled with the high edge density of graphene. Graphene edges provide significantly higher charge density and reactivity than the basal plane, but they are difficult to arrange in a three-dimensional, high volume-density geometry. CNTs are readily aligned in a high density geometry (i.e., a vertically aligned forest) but lack high charge density surfaces—the sidewalls of the CNTs are similar to the basal plane of graphene and exhibit low charge density except where edge defects exist. Depositing a high density of graphene foliates along the length of aligned CNTs can significantly increase the total charge capacity per unit of nominal area as compared to other carbon nanostructures.

Nitrogen-doped Carbon Nanotubes

Nitrogen doped carbon nanotubes (N-CNTs) can be produced through five main methods, chemical vapor deposition, high-temperature and high-pressure reactions, gas-solid reaction of amorphous carbon with NH_3 at high temperature, solid reaction, and solvothermal synthesis.

N-CNTs can also be prepared by a CVD method of pyrolyzing melamine under Ar at elevated temperatures of 800–980 °C. However synthesis by CVD of melamine results in the formation of bamboo-structured CNTs. XPS spectra of grown N-CNTs reveal nitrogen in five main components, pyridinic nitrogen, pyrrolic nitrogen, quaternary nitrogen, and nitrogen oxides. Furthermore, synthesis temperature affects the type of nitrogen configuration.

Nitrogen doping plays a pivotal role in lithium storage, as it creates defects in the CNT walls allowing for Li ions to diffuse into interwall space. It also increases capacity by providing more favorable bind of N-doped sites. N-CNTs are also much more reactive to metal oxide nanoparticle deposition which can further enhance storage capacity, especially in anode materials for Li-ion batteries. However boron-doped nanotubes have been shown to make batteries with triple capacity.

Peapod

A carbon peapod is a novel hybrid carbon material which traps fullerene inside a carbon nanotube. It can possess interesting magnetic properties with heating and irradiation. It can also be applied as an oscillator during theoretical investigations and predictions.

Cup-stacked Carbon Nanotubes

Cup-stacked carbon nanotubes (CSCNTs) differ from other quasi-1D carbon structures, which normally behave as quasi-metallic conductors of electrons. CSCNTs exhibit semiconducting behaviors due to the stacking microstructure of graphene layers.

Extreme Carbon Nanotubes

The observation of the *longest* carbon nanotubes grown so far are over 1/2 m (550 mm long) was reported in 2013. These nanotubes were grown on Si substrates using an improved chemical vapor deposition (CVD) method and represent electrically uniform arrays of single-walled carbon nanotubes.

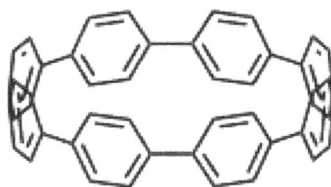

Cycloparaphenylene

The *shortest* carbon nanotube is the organic compound cycloparaphenylene, which was synthesized in early 2009.

The *thinnest* carbon nanotube is the armchair (2,2) CNT with a diameter of 0.3 nm. This nanotube was grown inside a multi-walled carbon nanotube. Assigning of carbon nanotube type was done by a combination of high-resolution transmission electron microscopy (HRTEM), Raman spectroscopy and density functional theory (DFT) calculations.

The *thinnest freestanding* single-walled carbon nanotube is about 0.43 nm in diameter. Researchers suggested that it can be either (5,1) or (4,2) SWCNT, but the exact type of carbon nanotube remains questionable. (3,3), (4,3) and (5,1) carbon nanotubes (all about 0.4 nm in diameter) were unambiguously identified using aberration-corrected high-resolution transmission electron microscopy inside double-walled CNTs.

The *highest density* of CNTs was achieved in 2013, grown on a conductive titanium-coated copper surface that was coated with co-catalysts cobalt and molybdenum at lower than typical temperatures of 450 °C. The tubes averaged a height of 380 nm and a mass density of 1.6 g cm^{-3}. The material showed ohmic conductivity (lowest resistance 22 kΩ).

Properties

Strength

Carbon nanotubes are the strongest and stiffest materials yet discovered in terms of tensile strength and elastic modulus respectively. This strength results from the covalent sp^2 bonds formed between the individual carbon atoms. In 2000, a multi-walled carbon nanotube was tested to have a tensile strength of 63 gigapascals (9,100,000 psi). (For illustration, this translates into the ability to endure tension of a weight equivalent to 6,422 kilograms-force (62,980 N; 14,160 lbf) on a cable with cross-section of 1 square millimetre (0.0016 sq in).)

Further studies, such as one conducted in 2008, revealed that individual CNT shells have strengths of up to ~100 gigapascals (15,000,000 psi), which is in agreement with quantum/atomistic models. Since carbon nanotubes have a low density for a solid of 1.3 to 1.4 g/cm³, its specific strength of up to 48,000 kN·m·kg⁻¹ is the best of known materials, compared to high-carbon steel's 154 kN·m·kg⁻¹.

Under excessive tensile strain, the tubes will undergo plastic deformation, which means the deformation is permanent. This deformation begins at strains of approximately 5% and can increase the maximum strain the tubes undergo before fracture by releasing strain energy.

Although the strength of individual CNT shells is extremely high, weak shear interactions between adjacent shells and tubes lead to significant reduction in the effective strength of multi-walled carbon nanotubes and carbon nanotube bundles down to only a few GPa. This limitation has been recently addressed by applying high-energy electron irradiation, which crosslinks inner shells and tubes, and effectively increases the strength of these materials to ~60 GPa for multi-walled carbon nanotubes and ~17 GPa for double-walled carbon nanotube bundles.

CNTs are not nearly as strong under compression. Because of their hollow structure and high aspect ratio, they tend to undergo buckling when placed under compressive, torsional, or bending stress.

Comparison of mechanical properties			
Material	Young's modulus (TPa)	Tensile strength (GPa)	Elongation at break (%)
SWNT[E]	~1 (from 1 to 5)	13−53	16
Armchair SWNT[T]	0.94	126.2	23.1
Zigzag SWNT[T]	0.94	94.5	15.6−17.5
Chiral SWNT	0.92		
MWNT[E]	0.2−0.8−0.95	11−63−150	
Stainless steel[E]	0.186−0.214	0.38−1.55	15−50
Kevlar−29&149[E]	0.06−0.18	3.6−3.8	~2

[E]Experimental observation; [T]Theoretical prediction

The above discussion referred to axial properties of the nanotube, whereas simple geometrical considerations suggest that carbon nanotubes should be much softer in the radial direction than along the tube axis. Indeed, TEM observation of radial elasticity suggested that even the van der Waals forces can deform two adjacent nanotubes. Nanoindentation experiments, performed by several groups on multiwalled carbon nanotubes and tapping/contact mode atomic force microscope measurements performed on single-walled carbon nanotubes, indicated a Young's modulus of the order of several GPa, confirming that CNTs are indeed rather soft in the radial direction.

Hardness

Standard single-walled carbon nanotubes can withstand a pressure up to 25 GPa without [plastic/permanent] deformation. They then undergo a transformation to superhard phase nanotubes. Maximum pressures measured using current experimental techniques are around 55 GPa. However, these new superhard phase nanotubes collapse at an even higher, albeit unknown, pressure.

The bulk modulus of superhard phase nanotubes is 462 to 546 GPa, even higher than that of diamond (420 GPa for single diamond crystal).

Wettability

The surface wettability of CNT is of importance for its applications in various settings. Although the intrinsic contact angle of graphite is around 90°, the contact angles of most as-synthesized CNT arrays are over 160°, exhibiting a superhydrophobic property. By applying a low voltage as low as 1.3V, the extreme water repellant surface can be switched into superhydrophilic.

Kinetic Properties

Multi-walled nanotubes are multiple concentric nanotubes precisely nested within one another. These exhibit a striking telescoping property whereby an inner nanotube core may slide, almost without friction, within its outer nanotube shell, thus creating an atomically perfect linear or rotational bearing. This is one of the first true examples of molecular nanotechnology, the precise positioning of atoms to create useful machines. Already, this property has been utilized to create the world's smallest rotational motor. Future applications such as a gigahertz mechanical oscillator are also envisioned.

Electrical Properties

Unlike graphene, which is a two-dimensional semimetal, carbon nanotubes are either metallic or semiconducting along the tubular axis. For a given (n,m) nanotube, if $n = m$, the nanotube is metallic; if $n - m$ is a multiple of 3, then the nanotube is semiconducting with a very small band gap, otherwise the nanotube is a moderate semiconductor. Thus all armchair ($n = m$) nanotubes are metallic, and nanotubes (6,4), (9,1), etc. are semiconducting. Carbon nanotubes are not semimetallic because the degenerate point (that point where the π [bonding] band meets the π* [anti-bonding] band, at which the energy goes to zero) is slightly shifted away from the K point in the Brillouin zone due to the curvature of the tube surface, casing hybridization between the σ* and π* anti-bonding bands, modifying the band dispersion.

The rule regarding metallic versus semiconductor behavior has exceptions, because curvature effects in small diameter tubes can strongly influence electrical properties. Thus, a (5,0) SWCNT that should be semiconducting in fact is metallic according to the calculations. Likewise, zigzag and chiral SWCNTs with small diameters that should be metallic have a finite gap (armchair nanotubes remain metallic). In theory, metallic nanotubes can carry an electric current density of 4×10^9 A/cm², which is more than 1,000 times greater than those of metals such as copper, where for copper interconnects current densities are limited by electromigration. Carbon nanotubes are thus being explored as conductivity enhancing components in composite materials and many groups are attempting to commercialize highly conducting electrical wire assembled from individual carbon nanotubes. There are significant challenges to be overcome, however, such as the much more resistive nanotube-to-nanotube junctions and impurities, all of which lower the electrical conductivity of the macroscopic nanotube wires by orders of magnitude, as compared to the conductivity of the individual nanotubes.

Because of its nanoscale cross-section, electrons propagate only along the tube's axis. As a result, carbon nanotubes are frequently referred to as one-dimensional conductors. The maximum electrical conductance of a single-walled carbon nanotube is $2G_0$, where $G_0 = 2e^2/h$ is the conductance of a single ballistic quantum channel.

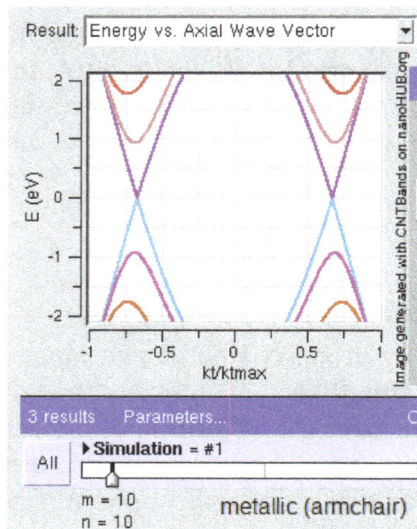

Band structures computed using tight binding approximation for (6,0) CNT (zigzag, metallic), (10,2) CNT (semiconducting) and (10,10) CNT (armchair, metallic).

Due to the role of the π-electron system in determining the electronic properties of graphene, doping in carbon nanotubes differs from that of bulk crystalline semiconductors from the same group of the periodic table (e.g. silicon). Graphitic substitution of carbon atoms in the nanotube wall by boron or nitrogen dopants leads to p-type and n-type behavior, respectively, as would be expected in silicon. However, some non-substitutional (intercalated or adsorbed) dopants introduced into a carbon nanotube, such as alkali metals as well as electron-rich metallocenes, result in n-type conduction because they donate electrons to the π-electron system of the nanotube. By contrast, π-electron acceptors such as $FeCl_3$ or electron-deficient metallocenes function as p-type dopants since they draw π-electrons away from the top of the valence band.

Intrinsic superconductivity has been reported, although other experiments found no evidence of this, leaving the claim a subject of debate.

Optical Properties

Thermal Properties

All nanotubes are expected to be very good thermal conductors along the tube, exhibiting a property known as "ballistic conduction", but good insulators lateral to the tube axis. Measurements show that a SWNT has a room-temperature thermal conductivity along its axis of about 3500 W·m⁻¹·K⁻¹; compare this to copper, a metal well known for its good thermal conductivity, which transmits 385 W·m⁻¹·K⁻¹. A SWNT has a room-temperature thermal conductivity across its axis (in the radial direction) of about 1.52 W·m⁻¹·K⁻¹, which is about as thermally conductive as soil. The temperature stability of carbon nanotubes is estimated to be up to 2800 °C in vacuum and about 750 °C in air.

Defects

As with any material, the existence of a crystallographic defect affects the material properties. Defects can occur in the form of atomic vacancies. High levels of such defects can lower the tensile strength by up to 85%. An important example is the Stone Wales defect, which creates a pentagon and heptagon pair by rearrangement of the bonds. Because of the very small structure of CNTs, the tensile strength of the tube is dependent on its weakest segment in a similar manner to a chain, where the strength of the weakest link becomes the maximum strength of the chain.

Crystallographic defects also affect the tube's electrical properties. A common result is lowered conductivity through the defective region of the tube. A defect in armchair-type tubes (which can conduct electricity) can cause the surrounding region to become semiconducting, and single monatomic vacancies induce magnetic properties.

Crystallographic defects strongly affect the tube's thermal properties. Such defects lead to phonon scattering, which in turn increases the relaxation rate of the phonons. This reduces the mean free path and reduces the thermal conductivity of nanotube structures. Phonon transport simulations indicate that substitutional defects such as nitrogen or boron will primarily lead to scattering of high-frequency optical phonons. However, larger-scale defects such as Stone Wales defects cause phonon scattering over a wide range of frequencies, leading to a greater reduction in thermal conductivity.

Safety and Health

The toxicity of carbon nanotubes has been an important question in nanotechnology. As of 2007, such research had just begun. The data is still fragmentary and subject to criticism. Preliminary results highlight the difficulties in evaluating the toxicity of this heterogeneous material. Parameters such as structure, size distribution, surface area, surface chemistry, surface charge, and agglomeration state as well as purity of the samples, have considerable impact on the reactivity of carbon nanotubes. However, available data clearly show that, under some conditions, nanotubes can cross membrane barriers, which suggests that, if raw materials reach the organs, they can induce harmful effects such as inflammatory and fibrotic reactions.

Synthesis

Techniques have been developed to produce nanotubes in sizable quantities, including arc discharge, laser ablation, high-pressure carbon monoxide disproportionation, and chemical vapor deposition (CVD). Most of these processes take place in a vacuum or with process gases. CVD growth of CNTs can occur in vacuum or at atmospheric pressure. Large quantities of nanotubes can be synthesized by these methods; advances in catalysis and continuous growth are making CNTs more commercially viable.

Chemical Modification

Carbon nanotubes can be functionalized to attain desired properties that can be used in a wide variety of applications. The two main methods of carbon nanotube functionalization are covalent and non-covalent modifications. Because of their hydrophobic nature, carbon nanotubes tend to agglomerate hindering their dispersion is solvents or viscous polymer melts. The resulting nanotube bundles or aggregates reduce the mechanical performance of the final composite. The surface of the carbon nanotubes can be modified to reduce the hydrophobicity and improve interfacial adhesion to a bulk polymer through chemical attachment.

Applications

Current

Current use and application of nanotubes has mostly been limited to the use of bulk nanotubes, which is a mass of rather unorganized fragments of nanotubes. Bulk nanotube materials may never achieve a tensile strength similar to that of individual tubes, but such composites may, nevertheless, yield strengths sufficient for many applications. Bulk carbon nanotubes have already been used as composite fibers in polymers to improve the mechanical, thermal and electrical properties of the bulk product.

- Easton-Bell Sports, Inc. have been in partnership with Zyvex Performance Materials, using CNT technology in a number of their bicycle components—including flat and riser handlebars, cranks, forks, seatposts, stems and aero bars.

- Zyvex Technologies has also built a 54' maritime vessel, the Piranha Unmanned Surface Vessel, as a technology demonstrator for what is possible using CNT technology. CNTs help improve the structural performance of the vessel, resulting in a lightweight 8,000 lb boat that can carry a payload of 15,000 lb over a range of 2,500 miles.

- Amroy Europe Oy manufactures Hybtonite carbon nanoepoxy resins where carbon nanotubes have been chemically activated to bond to epoxy, resulting in a composite material that is 20% to 30% stronger than other composite materials. It has been used for wind turbines, marine paints and variety of sports gear such as skis, ice hockey sticks, baseball bats, hunting arrows, and surfboards.

Other current applications include:

- tips for atomic force microscope probes

- in tissue engineering, carbon nanotubes can act as scaffolding for bone growth

There is also ongoing research in using carbon nanotubes as a scaffold for diverse microfabrication techniques.

Potential

The strength and flexibility of carbon nanotubes makes them of potential use in controlling other nanoscale structures, which suggests they will have an important role in nanotechnology engineering. The highest tensile strength of an individual multi-walled carbon nanotube has been tested to be 63 GPa. Carbon nanotubes were found in Damascus steel from the 17th century, possibly helping to account for the legendary strength of the swords made of it. Recently, several studies have highlighted the prospect of using carbon nanotubes as building blocks to fabricate three-dimensional macroscopic (>1mm in all three dimensions) all-carbon devices. Lalwani et al. have reported a novel radical initiated thermal crosslinking method to fabricated macroscopic, free-standing, porous, all-carbon scaffolds using single- and multi-walled carbon nanotubes as building blocks. These scaffolds possess macro-, micro-, and nano- structured pores and the porosity can be tailored for specific applications. These 3D all-carbon scaffolds/architectures maybe used for the fabrication of the next generation of energy storage, supercapacitors, field emission transistors, high-performance catalysis, photovoltaics, and biomedical devices and implants.

Discovery

The true identity of the discoverers of carbon nanotubes is a subject of some controversy. For years, scientists assumed that Sumio Iijima of NEC had discovered carbon nanotubes in 1991. He published a paper describing his discovery which initiated a flurry of excitement and could be credited by inspiring the many scientists now studying applications of carbon nanotubes. Though Iijima has been given much of the credit for discovering carbon nanotubes, it turns out that the timeline of carbon nanotubes goes back much further than 1991. In 1952 L. V. Radushkevich and V. M. Lukyanovich published clear images of 50 nanometer diameter tubes made of carbon in the Soviet *Journal of Physical Chemistry*. This discovery was largely unnoticed, as the article was published in Russian, and Western scientists' access to Soviet press was limited during the Cold War. Before they came to be known as carbon nanotubes, in 1976, Morinobu Endo of CNRS observed hollow tubes of rolled up graphite sheets synthesised by a chemical vapour-growth technique. The first specimens observed would later come to be known as single-walled carbon nanotubes (SWNTs). The three scientists have been the first ones to show images of a nanotube with a solitary graphene wall.

Endo, in his early review of vapor-phase-grown carbon fibers (VPCF), also reminded us that he had observed a hollow tube, linearly extended with parallel carbon layer faces near the fiber core. This appears to be the observation of multi-walled carbon nanotubes at the center of the fiber. The mass-produced MWCNTs today are strongly related to the VPGCF developed by Endo. In fact, they call it the "Endo-process", out of respect for his early work and patents.

In 1979, John Abrahamson presented evidence of carbon nanotubes at the 14th Biennial Conference of Carbon at Pennsylvania State University. The conference paper described carbon nanotubes as carbon fibers that were produced on carbon anodes during arc discharge. A characterization of these fibers was given as well as hypotheses for their growth in a nitrogen atmosphere at low pressures.

In 1981, a group of Soviet scientists published the results of chemical and structural characterization of carbon nanoparticles produced by a thermocatalytical disproportionation of carbon monoxide. Using TEM images and XRD patterns, the authors suggested that their "carbon multi-layer tubular crystals" were formed by rolling graphene layers into cylinders. They speculated that by rolling graphene layers into a cylinder, many different arrangements of graphene hexagonal nets are possible. They suggested two possibilities of such arrangements: circular arrangement (armchair nanotube) and a spiral, helical arrangement (chiral tube).

In 1987, Howard G. Tennent of Hyperion Catalysis was issued a U.S. patent for the production of "cylindrical discrete carbon fibrils" with a "constant diameter between about 3.5 and about 70 nanometers..., length 10^2 times the diameter, and an outer region of multiple essentially continuous layers of ordered carbon atoms and a distinct inner core...."

Iijima's discovery of multi-walled carbon nanotubes in the insoluble material of arc-burned graphite rods in 1991 and Mintmire, Dunlap, and White's independent prediction that if single-walled carbon nanotubes could be made, then they would exhibit remarkable conducting properties helped create the initial buzz that is now associated with carbon nanotubes. Nanotube research accelerated greatly following the independent discoveries by Bethune at IBM and Iijima at NEC of *single-walled* carbon nanotubes and methods to specifically produce them by adding transition-metal catalysts to the carbon in an arc discharge. The arc discharge technique was well-known to produce the famed Buckminster fullerene on a preparative scale, and these results appeared to extend the run of accidental discoveries relating to fullerenes. The discovery of nanotubes remains a contentious issue. Many believe that Iijima's report in 1991 is of particular importance because it brought carbon nanotubes into the awareness of the scientific community as a whole.

Silver Nanoparticle

Silver nanoparticles are nanoparticles of silver of between 1 nm and 100 nm in size. While frequently described as being 'silver' some are composed of a large percentage of silver oxide due to their large ratio of surface-to-bulk silver atoms. Numerous shapes of nanoparticles can be constructed depending on the application at hand. Commonly used are spherical silver nanoparticles but diamond, octagonal and thin sheets are also popular.

Their extremely large surface area permits the coordination of a vast number of ligands. The properties of silver nanoparticles applicable to human treatments are under investigation in laboratory and animal studies, assessing potential efficacy, toxicity, and costs.

Synthesis

Methods

Silver nanoparticles can be synthesized by several methods. They can be divided into three broad categories: wet chemistry, ion implantation, or biogenic synthesis.

Wet Chemistry

Several wet chemical methods have been developed to form silver nanoparticles. Each of these methods use the same basic mechanism: a silver ion complex is reduced to colloidal silver in the presence of a reducing agent, which oxidizes and gives back the missing electron, so that the silver ion is reduced back to its metallic state. The newly formed silver nanoparticles can then also be stabilized with a stabilizing/capping agent, which stabilizes the energy of the surface of the newly formed nanoparticle to prevent them from aggregate to larger particles. These methods include the use of reducing sugars, citrate reduction, reduction via sodium borohydride, the silver mirror reaction, the polyol process, seed-mediated growth, and light-mediated growth. Each of these methods, or combination of methods, will offer differing degrees of control over the size distribution as well as distributions of geometric arrangements of the nanoparticle. A new, very promising wet-chemical technique was found by Elsupikhe et al. (2015). They have developed a green ultrasonically-assisted synthesis. Under ultrasound treatment, silver nanoparticles (AgNP) are synthesized with κ-carrageenan as a natural stabilizer. The reaction is performed at ambient temperature and produces silver nanoparticles with fcc crystal structure without impurities. The concentration of κ-carrageenan is used to influence particle size distribution of the AgNPs.

Reduction with Reducing Sugars

Reducing sugars includes glucose, fructose, maltose, maltodextrin, etc. but not sucrose, and is the simpliest method to reduce silver ions back to silver nanoparticles. And if the pH is over 7 the glucose molucule in the reducing sugar will "open up" its ring structure and attach to the surface of the nanoparticles, and thereby act as a weak stabilizing agent as well.

Citrate Reduction

An early, and very common, method for synthesizing silver nanoparticles is citrate reduction. Citrate reduction involves the reduction of silver nitrate, $AgNO_3$, to colloidal silver using trisodium citrate, $Na_3C_6H_5O_7$ at elevated temperature (~100 °C). In this synthesis, the citrate ion acts as both the reducing and capping agent. This method is useful due to its relative ease and short reaction time. However, the silver particles formed may exhibit broad size distributions and/or form several different particle geometries simultaneously.

Reduction Via Sodium Borohydride

A similar method is the reduction of silver nitrate using sodium borohydride, NaBH4. Sodium borohydride is a stronger reducing agent than citrate, but it is not thought to participate in surface stabilization. Thus, a separate capping molecule must be used in addition to the sodium borohydride. Examples of stabilizers used include citrate, poly(vinyl pyrrolidone) (PVP), bovine serum albumin, and cetyltrimethylammonium bromide (CTAB).

Polyol Process

The polyol process is a particularly useful method because it yields a high degree of control over both the size and geometry of the resulting nanoparticles. In general, the polyol synthesis begins with the heating of a polyol compound such as ethylene glycol, 1,5-pentanediol, or 1,2-propylene glycol7. An Ag+ species and a capping agent are added (although the polyol itself is also often the capping agent). The Ag+ species is then reduced by the polyol to colloidal nanoparticles. The polyol process is highly sensitive to reaction conditions such as temperature, chemical environment, and concentration of substrates.(add refs again here) Therefore, by changing these variables, various sizes and geometries can be selected for such as quasi-spheres, pyramids, spheres, and wires. Further study has examined the mechanism for this process as well as resulting geometries under various reaction conditions in greater detail.

Seed-mediated Growth

Seed-mediated growth is the process in which seed nanoparticles, formed from a reduction of Ag+ to colloidal silver by any method, are placed in a growth solution. The growth solution contains an additional silver species that deposits more silver atoms onto the original seeds. Several experimental factors are manipulated to achieve size and geometry control in this method such as concentration of the substrates and the capping agent used. Parameters controlling the size and geometry of the particles is described in greater detail in various studies.

Light-mediated Growth

Light-mediated syntheses have also been explored where light can promote formation of various silver nanoparticle morphologies.

Silver Mirror Reaction

The silver mirror reaction involves the conversion of silver nitrate to $Ag(NH_3)OH$. $Ag(NH_3)OH$ is subsequently reduced into colloidal silver using an aldehyde containing molecule such as a sugar. The silver mirror reaction is as follows:

$$2(Ag(NH_3)_2)^+ + RCHO + 2OH^- \; RCOOH + 2Ag + 4NH_3$$

The size and shape of the nanoparticles produced are difficult to control and often have wide distributions. However, this method is often used to apply thin coatings of silver particles onto surfaces and further study into producing more uniformly sized nanoparticles is being done.

Ion Implantation

Ion implantation has been used to create silver nanoparticles embedded in glass, polyurethane, silicone, polyethylene, and poly(methyl methacrylate). Particles are embedded in the substrate by means of bombardment at high accelerating voltages. At a fixed current density of the ion beam up to a certain value, the size of the embedded silver nanoparticles has been found to be monodisperse within the population, after which only an increase in the ion concentration is observed. A further increase in the ion beam dose has been found to reduce both the nanoparticle size and density in the target substrate, whereas an ion beam operating at a high accelerating voltage with a gradually increasing current density has been found to result in a gradual increase in the nanoparticle size. There are a few competing mechanisms which may result in the decrease in nanoparticle size; destruction of NPs upon collision, sputtering of the sample surface, particle fusion upon heating and dissociation.

The formation of embedded nanoparticles is complex, and all of the controlling parameters and factors have not yet been investigated. Computer simulation is still difficult as it involves processes of diffusion and clustering, however it can be broken down into a few different sub-processes such as implantation, diffusion, and growth. Upon implantation, silver ions will reach different depths within the substrate which approaches a Gaussian distribution with the mean centered at X depth. High temperature conditions during the initial stages of implantation will increase the impurity diffusion in the substrate and as a result limit the impinging ion saturation, which is required for nanoparticle nucleation. Both the implant temperature and ion beam current density are crucial to control in order to obtain a monodisperse nanoparticle size and depth distribution. A low current density may be used to counter the thermal agitation from the ion beam and a buildup of surface charge. After implantation on the surface, the beam currents may be raised as the surface conductivity will increase. The rate at which impurities diffuse drops quickly after the formation of the nanoparticles, which act as a mobile ion trap. This suggests that the beginning of the implantation process is critical for control of the spacing and depth of the resulting nanoparticles, as well as control of the substrate temperature and ion beam density. The presence and nature of these particles can be analyzed using numerous spectroscopy and microscopy instruments. Nanoparticles synthesized in the substrate exhibit surface plasmon resonances as evidenced by characteristic absorption bands; these features undergo spectral shifts depending on the nanoparticle size and surface asperities, however the optical properties also strongly depend on the substrate material of the composite.

Biogenic Synthesis

The biological synthesis of nanoparticles has provided a means for improved techniques compared to the traditional methods that call for the use of harmful reducing agents like sodium borohydride. Many of these methods could improve their environmental footprint by replacing these relatively strong reducing agents. The problems with the chemical production of silver nanoparticles is usually involves high cost and the longevity of the particles is short lived due to aggregation. The harshness of standard chemical methods has sparked the use of using biological organisms to reduce silver ions in solution into colloidal nanoparticles.

In addition, precise control over shape and size is vital during nanoparticle synthesis since the NPs therapeutic properties are intimately dependent on such factors. Hence, the primary focus of research in biogenic synthesis is in developing methods that consistently reproduce NPs with precise properties.

Use of Fungi and Bacteria

A general representation of the synthesis and applications of biogenically synthesized silver nanoparticles using plant extract.

Bacterial and fungal synthesis of nanoparticles is practical because bacteria and fungi are easy to handle and can be modified genetically with ease. This provides a means to develop biomolecules that can synthesize AgNPs of varying shapes and sizes in high yield, which is at the forefront of current challenges in nanoparticle synthesis. Fungal strains such as Verticillium and bacterial strains such as K. pneumoniae can be used in the synthesis of silver nanoparticles. When the fungus/bacteria is added to solution, protein biomass is released into the solution. Electron donating residues such as tryptophan and tyrosine reduce silver ions in solution contributed by silver nitrate. These methods have been found to effectively create stable monodisperse nanoparticles without the use of harmful reducing agents.

A method has been found of reducing silver ions by the introduction of the fungus *Fusarium oxysporum*. The nanoparticles formed in this method have a size range between 5 and 15 nm and consist of silver hydrosol. The reduction of the silver nanoparticles is thought to come from an enzymatic process and silver nanoparticles produced are extremely stable due to interactions with proteins that are excreted by the fungi.

Bacterium found in silver mines, Pseudomonas stutzeri AG259, were able to construct silver particles in the shapes of triangles and hexagons. The size of these nanoparticles had a large range in size and some of them reached sizes larger than the usual nanoscale with a size of 200 nm. The silver nanoparticles were found in the organic matrix of the bacteria.

Lactic acid producing bacteria have been used to produce silver nanoparticles. The bacteria *Lactobacillus* spp., *Pediococcus pentosaceus*, *Enteroccus faeciumI*, and *Lactococcus garvieae* have been found to be able to reduce silver ions into silver nanoparticles. The production of the nanoparticles takes place in the cell from the interactions between the silver ions and the organic compounds of the cell. It was found that the bacterium *Lactobacillus fermentum* created the smallest silver nanoparticles with an average size of 11.2 nm. It was also found that this bacterium produced the nanoparticles with the smallest size distribution and the nanoparticles were found mostly on the outside of the cells. It was also found that there was an increase in the pH increased the rate of which the nanoparticles were produced and the amount of particles produced.

Use of Plants

The reduction of silver ions into silver nanoparticles has also been achieved using geranium leaves. It has been found that adding geranium leaf extract to silver nitrate solutions causes there silver ions to be quickly reduced and that the nanoparticles produced are particularly stable. The silver nanoparticles produced in solution had a size range between 16 and 40 nm.

In another study different plant leaf extracts were used to reduce silver ions. It was found that out of Camellia sinensis (green tea), pine, persimmon, ginko, magnolia, and platanus that the magnolia leaf extract was the best at creating silver nanoparticles. This method created particles with a disperse size range of 15 to 500 nm, but it was also found that the particle size could be controlled by varying the reaction temperature. The speed at which the ions were reduced by the magnolia leaf extract was comparable to those of using chemicals to reduce.

The use of plants, microbes, and fungi in the production of silver nanoparticles is leading the way to more environmentally sound production of silver nanoparticles.

Products and Functionalization

Synthetic protocols for silver nanoparticle production can be modified to produce silver nanoparticles with non-spherical geometries and also to functionalize nanoparticles with different materials, such as silica. Creating silver nanoparticles of different shapes and surface coatings allows for greater control over their size-specific properties.

Anisotropic Structures

Silver nanoparticles can be synthesized in a variety of non-spherical (anisotropic) shapes. Because silver, like other noble metals, exhibits a size and shape dependent optical effect known as localized surface plasmon resonance (LSPR) at the nanoscale, the ability to synthesize Ag nanoparticles in different shapes vastly increases the ability to tune their optical behavior. For example, the wavelength at which LSPR occurs for a nanoparticle of one morphology (e.g. a sphere) will be different if that sphere is changed into a different shape. This shape dependence allows a silver nanoparticle to experience optical enhancement at a range of different wavelengths, even by keeping the size relatively constant, just by changing its shape. The applications of this shape-exploited expansion of optical behavior range from developing more sensitive biosensors to increasing the longevity of textiles.

Triangular Nanoprisms

Triangular shaped nanoparticles are a canonical type of anisotropic morphology studied for both gold and silver.

Though many different techniques for silver nanoprism synthesis exist, several methods employ a seed-mediated approach, which involves first synthesizing small (3-5 nm diameter) silver nanoparticles that offer a template for shape-directed growth into triangular nanostructures.

The silver seeds are synthesized by mixing silver nitrate and sodium citrate in aqueous solution and then rapidly adding sodium borohydride. Additional silver nitrate is added to the seed solution at low temperature, and the prisms are grown by slowly reducing the excess silver nitrate using ascorbic acid.

With the seed-mediated approach to silver nanoprism synthesis, selectivity of one shape over another can in part be controlled by the capping ligand. Using essentially the same procedure above but changing citrate to poly (vinyl pyrrolidone) (PVP) yields cube and rod-shaped nanostructures instead of triangular nanoprisms.

In addition to the seed mediated technique, silver nanoprisms can also be synthesized using a photo-mediated approach, in which preexisting spherical silver nanoparticles are transformed into triangular nanoprisms simply by exposing the reaction mixture to high intensities of light.

Nanocubes

Silver nanocubes can be synthesized using ethylene glycol as a reducing agent and PVP as a capping agent, in a polyol synthesis reaction (vide supra). A typical synthesis using these reagents involves adding fresh silver nitrate and PVP to a solution of ethylene glycol heated at 140 °C.

This procedure can actually be modified to produce another anisotropic silver nanostructure, nanowires, by just allowing the silver nitrate solution to age before using it in the synthesis. By allowing the silver nitrate solution to age, the initial nanostructure formed during the synthesis is slightly different than that obtained with fresh silver nitrate, which influences the growth process, and therefore, the morphology of the final product.

Coating with Silica

In this method, polyvinylpyrrolidone (PVP) is dissolved in water by sonication and mixed with silver colloid particles. Active stirring ensures the PVP has adsorbed to the nanoparticle surface. Centrifuging separates the PVP coated nanoparticles which are then transferred to a solution of ethanol to be centrifuged further and placed in a solution of ammonia, ethanol and $Si(OEt_4)$ (TES). Stirring for twelve hours results in the silica shell being formed consisting of a surrounding layer of silicon oxide with an ether linkage available to add functionality. Varying the amount of TES allows for different thicknesses of shells formed. This technique is popular due to the ability to add a variety of functionality to the exposed silica surface.

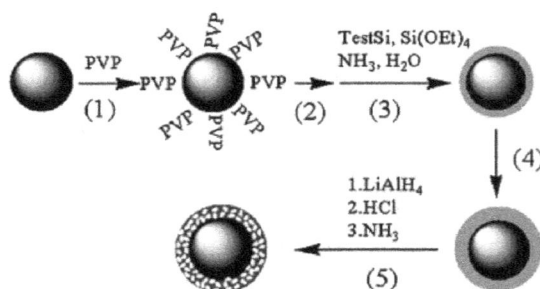

General procedure for coating colloid particles in silica. First PVP is absorbed onto the colloidal surface.
These particles are put into a solution of ammonia in ethanol. the particle then
begins to grow by addition of Si(OET4).

Use

Catalysis

Using silver nanoparticles for catalysis has been gaining attention in recent years. Although the most common applications are for medicinal or antibacterial purposes, silver nanoparticles have been demonstrated to show catalytic redox properties for dyes, benzene, carbon monoxide, and likely other compounds.

NOTE: This paragraph is a general description of nanoparticle properties for catalysis; it is not exclusive to silver nanoparticles. The size of a nanoparticle greatly determines the properties that it exhibits due to various quantum effects. Additionally, the chemical environment of the nanoparticle plays a large role on the catalytic properties. With this in mind, it is important to note that heterogeneous catalysis takes place by adsorption of the reactant species to the catalytic substrate. When polymers, complex ligands, or surfactants are used to prevent coalescence of the nanoparticles, the catalytic ability is frequently hindered due to reduced adsorption ability. However, these compounds can also be used in such a way that the chemical environment enhances the catalytic ability.

Supported on Silica Spheres – Reduction of Dyes

Silver nanoparticles have been synthesized on a support of inert silica spheres. The support plays virtually no role in the catalytic ability and serves as a method of preventing coalescence of the silver nanoparticles in colloidal solution. Thus, the silver nanoparticles were stabilized and it was possible to demonstrate the ability of them to serve as an electron relay for the reduction of dyes by sodium borohydride. Without the silver nanoparticle catalyst, virtually no reaction occurs between sodium borohydride and the various dyes: methylene blue, eosin, and rose Bengal.

Mesoporous Aerogel – Selective Oxidation of Benzene

Silver nanoparticles supported on aerogel are advantageous due to the higher number of active sites. The highest selectivity for oxidation of benzene to phenol was observed at low weight percent of silver in the aerogel matrix (1% Ag). This better selectivity is believed to be a result of the higher monodispersity within the aerogel matrix of the 1% Ag sample. Each weight percent solution formed different sized particles with a different width of size range.

Silver Alloy – Synergistic Oxidation of Carbon Monoxide

Au-Ag alloy nanoparticles have been shown to have a synergistic effect on the oxidation of carbon monoxide (CO). On its own, each pure-metal nanoparticle shows very poor catalytic activity for CO oxidation; together, the catalytic properties are greatly enhanced. It is proposed that the gold acts as a strong binding agent for the oxygen atom and the silver serves as a strong oxidizing catalyst, although the exact mechanism is still not completely understood. When synthesized in an Au/Ag ratio from 3:1 to 10:1, the alloyed nanoparticles showed complete conversion when 1% CO was fed in air at ambient temperature. Interestingly, the size of the alloyed particles did not play a big role in the catalytic ability. It is well known that gold nanoparticles only show catalytic properties for CO when they are ~3 nm in size, but alloyed particles up to 30 nm demonstrated excellent catalytic activity – catalytic activity better than that of gold nanoparticles on active support such as TiO_2, Fe_2O_3, etc.

Light-enhanced

Plasmonic effects have been studied quite extensively. Until recently, there have not been studies investigating the oxidative catalytic enhancement of a nanostructure via excitation of its surface plasmon resonance. The defining feature for enhancing the oxidative catalytic ability has been identified as the ability to convert a beam of light into the form of energetic electrons that can be transferred to adsorbed molecules. The implication of such a feature is that photochemical reactions can be driven by low-intensity continuous light can be coupled with thermal energy.

The coupling of low-intensity continuous light and thermal energy has been performed with silver nanocubes. The important feature of silver nanostructures that are enabling for photocatalysis is their nature to create resonant surface plasmons from light in the visible range.

The addition of light enhancement enabled the particles to perform to the same degree as particles that were heated up to 40K greater. This is a profound finding when noting that a reduction in temperature of 25K can increase the catalyst lifetime by nearly tenfold, when comparing the photothermal and thermal process.

Biological Research

Researchers have explored the use of silver nanoparticles as carriers for delivering various payloads such as small drug molecules or large biomolecules to specific targets. Once the AgNP has had sufficient time to reach its target, release of the payload could potentially be triggered by an internal or external stimulus. The targeting and accumulation of nanoparticles may provide high payload concentrations at specific target sites and could minimize side effects.

Chemotherapy

The introduction of nanotechnology into medicine is expected to advance diagnostic cancer imaging and the standards for therapeutic drug design. Nanotechnology may uncover insight about the structure, function and organizational level of the biosystem at the nanoscale.

Silver nanoparticles can undergo coating techniques that offer a uniform functionalized surface to which substrates can be added. When the nanoparticle is coated, for example, in silica the surface exists as silicic acid. Substrates can thus be added through stable ether and ester linkages that are not degraded immediately by natural metabolic enzymes. Recent chemotherapeutic applications have designed anti cancer drugs with a photo cleavable linker, such as an ortho-nitrobenzyl bridge, attaching it to the substrate on the nanoparticle surface. The low toxicity nanoparticle complex can remain viable under metabolic attack for the time necessary to be distributed throughout the bodies systems. If a cancerous tumor is being targeted for treatment, ultraviolet light can be introduced over the tumor region. The electromagnetic energy of the light causes the photo responsive linker to break between the drug and the nanoparticle substrate. The drug is now cleaved and released in an unaltered active form to act on the cancerous tumor cells. Advantages anticipated for this method is that the drug is transported without highly toxic compounds, the drug is released without harmful radiation or relying on a specific chemical reaction to occur and the drug can be selectively released at a target tissue.

A second approach is to attach a chemotherapeutic drug directly to the functionalized surface of the silver nanoparticle combined with a nucelophilic species to undergo a displacement reaction. For example, once the nanoparticle drug complex enters or is in the vicinity of the target tissue or cells, a glutathione monoester can be administered to the site. The nucleophilic ester oxygen will attach to the functionalized surface of the nanoparticle through a new ester linkage while the drug is released to its surroundings. The drug is now active and can exert its biological function on the cells immediate to its surroundings limiting non-desirable interactions with other tissues.

Multiple Drug Resistance

A major cause for the ineffectiveness of current chemotherapy treatments is multiple drug resistance which can arise from several mechanisms.

Nanoparticles can provide a means to overcome MDR. In general, when using a targeting agent to deliver nanocarriers to cancer cells, it is imperative that the agent binds with high selectivity to molecules that are uniquely expressed on the cell surface. Hence NPs can be designed with proteins that specifically detect drug resistant cells with overexpressed transporter proteins on their surface. A pitfall of the commonly used nano-drug delivery systems is that free drugs that are released from the nanocarriers into the cytosol get exposed to the MDR transporters once again, and are exported. To solve this, 8 nm nano crystalline silver particles were modified by the addition of trans-activating transcriptional activator (TAT), derived from the HIV-1 virus, which acts as a cell penetrating peptide (CPP). Generally, AgNP effectiveness is limited due to the lack of efficient cellular uptake; however, CPP-modification has become one of the most efficient methods for improving intracellular delivery of nanoparticles. Once ingested, the export of the AgNP is prevented based on a size exclusion. The concept is simple: the nanoparticles are too large to be effluxed by the MDR transporters, because the efflux function is strictly subjected to the size of its substrates, which is generally limited to a range of 300-2000 Da. Thereby the nanoparticulates remain insusceptible to the efflux, providing a means to accumulate in high concentrations.

Antimicrobial

Introduction of silver into bacterial cells induces a high degree of structural and morphological changes, which can lead to cell death. As the silver nano particles come in contact with the bacteria, they adhere to the cell wall and cell membrane. Once bound, some of the silver passes through to the inside, and interacts with phosphate-containing compounds like DNA and RNA, while another portion adheres to the sulphur-containing proteins on the membrane. The silver-sulphur interactions at the membrane cause the cell wall to undergo structural changes, like the formation of pits and pores. Through these pores, cellular components are released into the extracellular fluid, simply due to the osmotic difference. Within the cell, the integration of silver creates a low molecular weight region where the DNA then condenses. Having DNA in a condensed state inhibits the cell's replication proteins contact with the DNA. Thus the introduction of silver nanoparticles inhibits replication and is sufficient to cause the death of the cell. Further increasing their effect, when silver comes in contact with fluids, it tends to ionize which increases the nanoparticles bactericidal activity. This has been correlated to the suppression of enzymes and inhibited expression of proteins that relate to the cell's ability to produce ATP.

Although it varies for every type of cell proposed, as their cell membrane composition varies greatly, It has been seen that in general, silver nano particles with an average size of 10 nm or less show electronic effects that greatly increase their bactericidal activity. This could also be partly due to the fact that as particle size decreases, reactivity increases due to the surface area to volume ratio increasing.

It has been noted that the introduction of silver nano particles has shown to have synergistic activity with common antibiotics already used today, such as; penicillin G, ampicillin, erythromycin, clindamycin, and vancomycin against E. coli and S. aureus. In medical equipment, it has been shown that silver nano particles drastically lower the bacterial count on devices used. However, the problem arises when the procedure is over and a new one must be done. In the process of washing the instruments a large portion of the silver nano particles become less effective due to the loss of silver ions. They are more commonly used in skin grafts for burn victims as the silver nano particles embedded with the graft provide better antimicrobial activity and result in significantly less scarring of the victim. They also show promising application as water treatment method to form clean potable water.

Other

Samsung has created and marketed a material called Silver Nano, that includes silver nanoparticles on the surfaces of household appliances.

Safety

Although silver nanoparticles are widely used in a variety of commercial products, there has only recently been a major effort to study their effects on human health. There have been several studies that describe the *in vitro* toxicity of silver nanoparticles to a variety of different organs, including the lung, liver, skin, brain, and reproductive organs. The mechanism of the toxicity of silver nanoparticles to human cells appears to be derived from oxidative stress and inflammation that is caused by the generation of reactive oxygen species (ROS) stimulated by either the Ag NPs, Ag ions, or both. For example, Park *et al.* showed that exposure of a mouse peritoneal macrophage cell line (RAW267.7) to silver nanoparticles decreased the cell viability in a concentration- and time-dependent manner. They further showed that the intracellular reduced glutathionine (GSH), which is a ROS scavenger, decreased to 81.4% of the control group of silver nanoparticles at 1.6 ppm.

Modes of Toxicity

Since silver nanoparticles undergo dissolution releasing silver ions, which is well-documented to have toxic effects, there have been several studies that have been conducted to determine whether the toxicity of silver nanoparticles is derived from the release of silver ions or from the nanoparticle itself. Several studies suggest that the toxicity of silver nanoparticles is attributed to their release of silver ions in cells as both silver nanoparticles and silver ions have been reported to have similar cytotoxicity. For example, In some cases it is reported that silver nanoparticles facilitate the release of toxic free silver ions in cells via a "Trojan-horse type mechanism," where the particle enters cells and is then ionized within the cell. However, there have been reports that suggest that a combination of silver nanoparticles and ions is responsible for the toxic effect of silver nanoparticles. Navarro *et al.* using cysteine ligands as a tool to measure the concentration of free silver in solution, determined that although initially silver ions were 18 times more likely to inhibit the photosynthesis of an algae, Chlamydomanas reinhardtii, but after 2 hours of incubation it was revealed that the algae containing silver nanoparticles were more toxic than just silver ions alone. Furthermore, there are studies that suggest that silver nanoparticles induce toxicity independent of free silver ions. For example, Asharani *et al.* compared phenotypic defects observed in zebrafish treated with silver nanoparticles and silver ions and determined that the phenotypic defects observed with silver nanoparticle treatment was not observed with silver ion-treated embryos, suggesting that the toxicity of silver nanoparticles are independent of silver ions.

Protein channels and nuclear membrane pores can often be in the size range of 9 nm to 10 nm in diameter. Small silver nanoparticles constructed of this size have the ability to not only pass through the membrane to interact with internal structures but also to be become lodged within the membrane. Silver nanoparticle depositions in the membrane can impact regulation of solutes, exchange of proteins and cell recognition. Exposure to silver nanoparticles has been associated with "inflammatory, oxidative, genotoxic, and cytotoxic consequences"; the silver particulates primarily accumulate in the liver. but have also been shown to be toxic in other organs including the brain. Nano-silver applied to tissue-cultured human cells leads to the formation of free radicals, raising concerns of potential health risks.

- Allergic reaction: There have been several studies conducted that show a precedence for allerginicity of silver nanoparticles.

- Argyria and staining: Ingested silver or silver compounds, including colloidal silver, can cause a condition called argyria, a discoloration of the skin and organs.In 2006, there was a case study of a 17-year-old man, who sustained burns to 30% of his body, and experienced a temporary bluish-grey hue after several days of treatment with Acticoat, a brand of wound dressing containing silver nanoparticles. Argyria is the deposition of silver in deep tissues, a condition that cannot happen on a temporary basis, raising the question of whether the cause of the man's discoloration was argyria or even a result of the silver treatment. Silver dressings are known to cause a "transient discoloration" that dissipates in 2–14 days, but not a permanent discoloration.

- Silzone heart valve: St. Jude Medical released a mechanical heart valve with a silver coated sewing cuff (coated using ion beam-assisted deposition) in 1997. The valve was designed to reduce the instances of endocarditis. The valve was approved for sale in Canada, Europe, the United States, and most other markets around the world. In a post-commercialization study, researchers showed that the valve prevented tissue ingrowth, created paravalvular leakage, valve loosening, and in the worst cases explantation. After 3 years on the market and 36,000 implants, St. Jude discontinued and voluntarily recalled the valve.

Hybrid Material

Hybrid materials are composites consisting of two constituents at the nanometer or molecular level. Commonly one of these compounds is inorganic and the other one organic in nature. Thus, they differ from traditional composites where the constituents are at the macroscopic (micrometer to millimeter) level. Mixing at the microscopic scale leads to a more homogeneous material that either show characteristics in between the two original phases or even new properties.

Introduction

Hybrid Materials in Nature

Many natural materials consist of inorganic and organic building blocks distributed on the nanoscale. In most cases the inorganic part provides mechanical strength and an overall structure to the natural objects while the organic part delivers bonding between the inorganic building blocks and/or the soft tissue. Typical examples of such materials are bone, or nacre.

Development of Hybrid Materials

The first hybrid materials were the paints made from inorganic and organic components that were used thousands of years ago. Rubber is an example of the use of inorganic materials as fillers for organic polymers. The sol–gel process developed in the 1930s was one of the major driving forces what has become the broad field of inorganic–organic hybrid materials.

Classification

Hybrid materials can be classified based on the possible interactions connecting the inorganic and organic species. *Class I* hybrid materials are those that show weak interactions between the two phases, such as van der Waals, hydrogen bonding or weak electrostatic interactions. *Class II* hybrid materials are those that show strong chemical interactions between the components such as covalent bonds.

Structural properties can also be used to distinguish between various hybrid materials. An organic moiety containing a functional group that allows the attachment to an inorganic network, e.g. a trialkoxysilane group, can act as a *network modifier* because in the final structure the inorganic network is only modified by the organic group. Phenyltrialkoxysilanes are an example for such compounds; they modify the silica network in the sol–gel process via the reaction of the trialkoxysilane group without supplying additional functional groups intended to undergo further chemical reactions to the material formed. If a reactive functional group is incorporated the system is called a *network functionalizer*. The situation is different if two or three of such anchor groups modify an organic segment; this leads to materials in which the inorganic group is afterwards an integral part of the hybrid network. The latter type of system is known as *network builder*

Blends are formed if no strong chemical interactions exist between the inorganic and organic building blocks. One example for such a material is the combination of inorganic clusters or particles with organic polymers lacking a strong (e.g. covalent) interaction between the components. In this case a material is formed that consists for example of an organic polymer with entrapped discrete inorganic moieties in which, depending on the functionalities of the components, for example weak crosslinking occurs by the entrapped inorganic units through physical interactions or the inorganic components are entrapped in a crosslinked polymer matrix. If an inorganic and an organic network interpenetrate each other without strong chemical interactions, so called interpenetrating networks (IPNs) are formed, which is for example the case if a sol–gel material is formed in presence of an organic polymer or vice versa. Both materials described belong to class I hybrids. Class II hybrids are formed when the discrete inorganic building blocks, e.g. clusters, are covalently bonded to the organic polymers or inorganic and organic polymers are covalently connected with each other.

Distinction between Nanocomposites and Hybrid Materials

The term nanocomposite is used if one of the structural units, either the organic or the inorganic, is in a defined size range of 1–100 nm. Large molecular building blocks for hybrid materials, such as large inorganic clusters, may be of the nanometer length scale. The term nanocomposites is generally used if discrete structural units in the respective size regime are used and the term hybrid materials is more often used if the inorganic units are formed in situ by molecular precursors, for example applying sol–gel reactions. Examples of discrete inorganic units for nanocomposites are nanoparticles, nanorods and carbon nanotubes. The biggest distinction between a nanocomposite and a hybrid is that a hybrid material possesses a property that does not exist in either of the parent components.<G. L. Drisko, C. Sanchez, Eur. J. Inorg. Chem. 2012, 2012, 5097>

Advantages of Hybrid Materials over Traditional Composites

- Inorganic clusters or nanoparticles with specific optical, electronic or magnetic properties can be incorporated in organic polymer matrices.

- Contrary to pure solid state inorganic materials that often require a high temperature treatment for their processing, hybrid materials show a more polymer-like handling, either because of their large organic content or because of the formation of crosslinked inorganic networks from small molecular precursors just like in polymerization reactions.

- Light scattering in homogeneous hybrid material can be avoided and therefore optical transparency of the resulting hybrid materials and nanocomposites can be achieved.

Synthesis

Two different approaches can be used for the formation of hybrid materials: Either well-defined preformed building blocks are applied that react with each other to form the final hybrid material in which the precursors still at least partially keep their original integrity or one or both structural units are formed from the precursors that are transformed into a new (network) structure. It is important that the interface between the inorganic and organic materials which has to be tailored to overcome serious problems in the preparation of hybrid materials. Different building blocks and approaches can be used for their preparation and these have to be adapted to bridge the differences of inorganic and organic materials.

Building Block Approach

Building blocks at least partially keep their molecular integrity throughout the material formation, which means that structural units that are present in these sources for materials formation can also be found in the final material. At the same time typical properties of these building blocks usually survive the matrix formation, which is not the case if material precursors are transferred into novel materials. Representative examples of such well-defined building blocks are modified inorganic clusters or nanoparticles with attached reactive organic groups.

Cluster compounds often consist of at least one functional group that allows an interaction with an organic matrix, for example by copolymerization. Depending on the number of groups that can interact, these building blocks are able to modify an organic matrix (one functional group) or form partially or fully crosslinked materials (more than one group). For instance, two reactive groups can lead to the formation of chain structures. If the building blocks contain at least three reactive groups they can be used without additional molecules for the formation of a crosslinked material.

Beside the molecular building blocks mentioned, nanosized building blocks, such as particles or nanorods, can also be used to form nanocomposites. The building block approach has one large advantage compared with the in situ formation of the inorganic or organic entities: because at least one structural unit (the building block) is well-defined and usually does not undergo significant structural changes during the matrix formation, better structure–property predictions are possible. Furthermore, the building blocks can be designed in such a way to give the best performance in the materials' formation, for example good solubility of inorganic compounds in organic monomers by surface groups showing a similar polarity as the monomers.

In recent years many building blocks have been synthesized and used for the preparation of hybrid materials. Chemists can design these compounds on a molecular scale with highly sophisticated methods and the resulting systems are used for the formation of functional hybrid materials. Many future applications, in particular in nanotechnology, focus on a bottom-up approach in which complex structures are hierarchically formed by these small building blocks. This idea is also one of the driving forces of the building block approach in hybrid materials.

In Situ Formation of The Components

The in situ formation of the hybrid materials is based on the chemical transformation of the precursors used throughout materials' preparation. Typically this is the case if organic polymers are formed but also if the sol–gel process is applied to produce the inorganic component. In these cases well-defined discrete molecules are transformed to multidimensional structures, which often show totally different properties from the original precursors. Generally simple, commercially available molecules are applied and the internal structure of the final material is determined by the composition of these precursors but also by the reaction conditions. Therefore control over the latter is a crucial step in this process. Changing one parameter can often lead to two very different materials. If, for example, the inorganic species is a silica derivative formed by the sol–gel process, the change from base to acid catalysis makes a large difference because base catalysis leads to a more particle-like microstructure while acid catalysis leads to a polymer-like microstructure. Hence, the final performance of the derived materials is strongly dependent on their processing and its optimization.

In Situ Formation of Inorganic Materials

Many of the classical inorganic solid state materials are formed using solid precursors and high temperature processes, which are often not compatible with the presence of organic groups because they are decomposed at elevated temperatures. Hence, these high temperature processes are not suitable for the in situ formation of hybrid materials. Reactions that are employed should have more the character of classical covalent bond formation in solutions. One of the most prominent processes which fulfill these demands is the sol–gel process. However, such rather low temperature processes often do not lead to the thermodynamically most stable structure but to kinetic products, which has some implications for the structures obtained. For example low temperature derived inorganic materials are often amorphous or crystallinity is only observed on a very small length scale, i.e. the nanometer range. An example of the latter is the formation of metal nanoparticles in organic or inorganic matrices by reduction of metal salts or organometallic precursors.

Some methods of in situ formation of inorganic materials are:

- Sol-gel process

- Nonhydrolytic sol–gel process

- Sol–gel reactions of non-silicates

Formation of Organic Polymers in Presence of Preformed Inorganic Materials

If the organic polymerization occurs in the presence of an inorganic material to form the hybrid material one has to distinguish between several possibilities to overcome the incompatibility of the two species. The inorganic material can either have no surface functionalization but the bare material surface; it can be modified with nonreactive organic groups (e.g. alkyl chains); or it can contain reactive surface groups such as polymerizable functionalities. Depending on these prerequisites the material can be pretreated, for example a pure inorganic surface can be treated with surfactants or silane coupling agents to make it compatible with the organic monomers, or functional monomers can be added that react with the surface of the inorganic material. If the inorganic component has nonreactive organic groups attached to its surface and it can be dissolved in a monomer which is subsequently polymerized, the resulting material after the organic polymerization, is a blend. In this case the inorganic component interact only weakly or not at all with the organic polymer; hence, a class I material is formed. Homogeneous materials are only obtained in this case if agglomeration of the inorganic components in the organic environment is prevented. This can be achieved if the interactions between the inorganic components and the monomers are better or at least the same as between the inorganic components. However, if no strong chemical interactions are formed, the long-term stability of a once homogeneous material is questionable because of diffusion effects in the resulting hybrid material. The stronger the respective interaction between the components, the more stable is the final material. The strongest interaction is achieved if class II materials are formed, for example with covalent interactions.

Hybrid Materials by Simultaneous Formation of both Components

Simultaneous formation of the inorganic and organic polymers can result in the most homogeneous type of interpenetrating networks. Usually the precursors for the sol–gel process are mixed with monomers for the organic polymerization and both processes are carried out at the same time with or without solvent. Applying this method, three processes are competing with each other:

(a)   the kinetics of the hydrolysis and condensation forming the inorganic phase,

(b)   the kinetics of the polymerization of the organic phase, and

(c)   the thermodynamics of the phase separation between the two phases.

By tailoring the kinetics of the two polymerizations in such a way that they occur simultaneously and rapidly enough, phase separation is avoided or minimized. Additional parameters such as attractive interactions between the two moieties, as described above can also be used to avoid phase separation.

One problem that also arises from the simultaneous formation of both networks is the sensitivity of many organic polymerization processes for sol–gel conditions or the composition of the materials formed. Ionic polymerizations, for example, often interact with the precursors or intermediates formed in the sol–gel process. Therefore, they are not usually applied in these reactions.

Applications

- Decorative coatings obtained by the embedding of organic dyes in hybrid coatings.

- Scratch-resistant coatings with hydrophobic or anti-fogging properties.

- Nanocomposite based devices for electronic and optoelectronic applications including light-emitting diodes, photodiodes, solar cells, gas sensors and field effect transistors.

- Fire retardant materials for construction industry.

- Nanocomposite based dental filling materials.

- Composite electrolyte materials for applications such as solid-state lithium batteries or supercapacitors.

- Proton conducting membranes used in fuel cells.

- Antistatic / anti-reflection coatings

- Corrosion protection

- Porous hybrid materials

Nanowire

A nanowire is a nanostructure, with the diameter of the order of a nanometer (10^{-9} meters). It can also be defined as the ratio of the length to width being greater than 1000. Alternatively, nanowires can be defined as structures that have a thickness or diameter constrained to tens of nanometers or less and an unconstrained length. At these scales, quantum mechanical effects are important — which coined the term "quantum wires". Many different types of nanowires exist, including superconducting (e.g., YBCO), metallic (e.g., Ni, Pt, Au), semiconducting (e.g., Si, InP, GaN, etc.), and insulating (e.g., SiO2, TiO2). Molecular nanowires are composed of repeating molecular units either organic (e.g. DNA) or inorganic (e.g. $Mo_6S_{9-x}I_x$).

Overview

Typical nanowires exhibit aspect ratios (length-to-width ratio) of 1000 or more. As such they are often referred to as one-dimensional (1-D) materials. Nanowires have many interesting properties that are not seen in bulk or 3-D (three-dimensional) materials. This is because electrons in nanowires are quantum confined laterally and thus occupy energy levels that are different from the traditional continuum of energy levels or bands found in bulk materials.

Crystalline 2×2-atom tin selenide nanowire grown inside a single-wall carbon nanotube (tube diameter ~1 nm).

Peculiar features of this quantum confinement exhibited by certain nanowires manifest themselves in discrete values of the electrical conductance. Such discrete values arise from a quantum mechanical restraint on the number of electrons that can travel through the wire at the nanometer scale. These discrete values are often referred to as the quantum of conductance and are integer multiples of

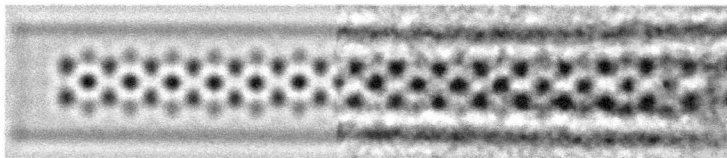

A noise-filtered HRTEM image of a HgTe extreme nanowire embedded down the central pore of a SWCNT. The image is also accompanied by a simulation of the crystal structure.

They are inverse of the well-known resistance unit h/e^2, which is roughly equal to 25812.8 ohms, and referred to as the von Klitzing constant R_K (after Klaus von Klitzing, the discoverer of exact quantization). Since 1990, a fixed conventional value R_{K-90} is accepted.

Examples of nanowires include inorganic molecular nanowires ($Mo_6S_{9-x}I_x$, $Li_2Mo_6Se_6$), which can have a diameter of 0.9 nm and be hundreds of micrometers long. Other important examples are based on semiconductors such as InP, Si, GaN, etc., dielectrics (e.g. SiO_2, TiO_2), or metals (e.g. Ni, Pt).

There are many applications where nanowires may become important in electronic, opto-electronic and nanoelectromechanical devices, as additives in advanced composites, for metallic interconnects in nanoscale quantum devices, as field-emitters and as leads for biomolecular nanosensors.

Synthesis of Nanowires

There are two basic approaches to synthesizing nanowires: top-down and bottom-up. A top-down approach reduces a large piece of material to small pieces, by various means such as lithography or electrophoresis. A bottom-up approach synthesizes the nanowire by combining constituent adatoms. Most synthesis techniques use a bottom-up approach.

An SEM image of epitaxial nanowire heterostructures grown from catalytic gold nanoparticles.

Nanowire production uses several common laboratory techniques, including suspension, electrochemical deposition, vapor deposition, and VLS growth. Ion track technology enables growing homogeneous and segmented nanowires down to 8 nm diameter.

Suspension

A suspended nanowire is a wire produced in a high-vacuum chamber held at the longitudinal extremities. Suspended nanowires can be produced by:

- The chemical etching of a larger wire

- The bombardment of a larger wire, typically with highly energetic ions

- Indenting the tip of a STM in the surface of a metal near its melting point, and then retracting it

VLS Growth

A common technique for creating a nanowire is vapor-liquid-solid method (VLS). This process can produce crystalline nanowires of some semiconductor materials. It uses a source material from either laser ablated particles or a feed gas such as silane.

VLS synthesis requires a catalyst. For nanowires, the best catalysts are liquid metal (such as gold) nanoclusters, which can either be self-assembled from a thin film by dewetting, or purchased in colloidal form and deposited on a substrate.

The source enters these nanoclusters and begins to saturate them. On reaching supersaturation, the source solidifies and grows outward from the nanocluster. Simply turning off the source can adjust the final length of the nanowire. Switching sources while still in the growth phase can create compound nanowires with super-lattices of alternating materials.

A single-step vapour phase reaction at elevated temperature synthesises inorganic nanowires such as $Mo_6S_{9-x}I_x$. From another point of view, such nanowires are cluster polymers.

Solution-phase Synthesis

Solution-phase synthesis refers to techniques that grow nanowires in solution. They can produce nanowires of many types of materials. Solution-phase synthesis has the advantage that it can produce very large quantities, compared to other methods. In one technique, the polyol synthesis, ethylene glycol is both solvent and reducing agent. This technique is particularly versatile at producing nanowires of lead, platinum, and silver.

The supercritical fluid-liquid-solid growth method can be used to synthesize semiconductor nanowires, e.g., Si and Ge. By using metal nanocrystals as seeds, Si and Ge organometallic precursors are fed into a reactor filled with a supercritical organic solvent, such as toluene. Thermolysis results in degradation of the precursor, allowing release of Si or Ge, and dissolution into the metal nanocrystals. As more of the semiconductor solute is added from the supercritical phase (due to a concentration gradient), a solid crystallite precipitates, and a nanowire grows uniaxially from the nanocrystal seed.

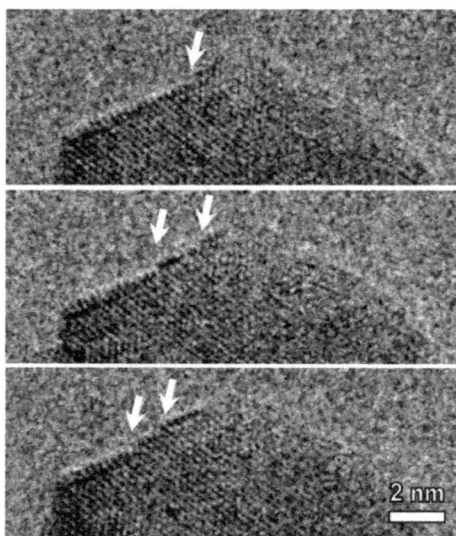

In situ observation of CuO nanowire growth

Non-catalytic Growth

Nanowires can be also grown without the help of catalysts, which gives an advantage of pure nanowires and minimizes the number of technological steps. The simplest methods to obtain metal oxide nanowires use ordinary heating of the metals, e.g. metal wire heated with battery, by Joule heating in air can be easily done at home. The vast majority of nanowire-formation mechanisms are explained through the use of catalytic nanoparticles, which drive the nanowire growth and are either added intentionally or generated during the growth. However the mechanisms for catalyst-free growth of nanowires (or whiskers) were known from 1950s. Spontaneous nanowire formation by non-catalytic methods were explained by the dislocation present in specific directions or the growth anisotropy of various crystal faces. More recently, after microscopy advancement, the nanowire growth driven by screw dislocations or twin boundaries were demonstrated. The picture on the right shows a single atomic layer growth on the tip of CuO nanowire, observed by in situ TEM microscopy during the non-catalytic synthesis of nanowire.

Physics of Nanowires

Conductivity of Nanowires

Several physical reasons predict that the conductivity of a nanowire will be much less than that of the corresponding bulk material. First, there is scattering from the wire boundaries, whose effect will be very significant whenever the wire width is below the free electron mean free path of the bulk material. In copper, for example, the mean free path is 40 nm. Copper nanowires less than 40 nm wide will shorten the mean free path to the wire width.

Nanowires also show other peculiar electrical properties due to their size. Unlike single wall carbon nanotubes, whose motion of electrons can fall under the regime of ballistic transport (meaning the electrons can travel freely from one electrode to the other), nanowire conductivity is strongly influenced by edge effects. The edge effects come from atoms that lay at the nanowire surface and are not fully bonded to neighboring atoms like the atoms within the bulk of the nanowire. The unbonded atoms are often a source of defects within the nanowire, and may cause the nanowire to conduct electricity more poorly than the bulk material. As a nanowire shrinks in size, the surface atoms become more numerous compared to the atoms within the nanowire, and edge effects become more important.

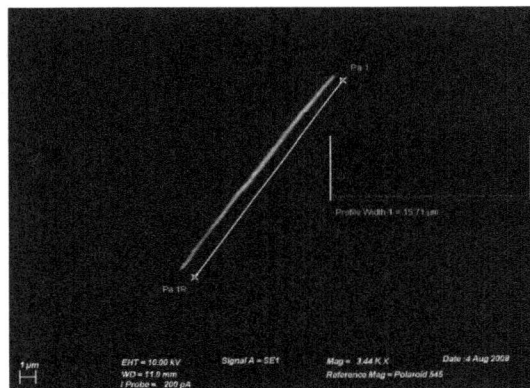

An SEM image of a 15 micrometer nickel wire.

Furthermore, the conductivity can undergo a quantization in energy: i.e. the energy of the electrons going through a nanowire can assume only discrete values, which are multiples of the conductance quantum $G = 2e^2/h$ (where e is the charge of the electron and h is the Planck constant.

The conductivity is hence described as the sum of the transport by separate *channels* of different quantized energy levels. The thinner the wire is, the smaller the number of channels available to the transport of electrons.

This quantization has been demonstrated by measuring the conductivity of a nanowire suspended between two electrodes while pulling it: as its diameter reduces, its conductivity decreases in a stepwise fashion and the plateaus correspond to multiples of G.

The quantization of conductivity is more pronounced in semiconductors like Si or GaAs than in metals, due to their lower electron density and lower effective mass. It can be observed in 25 nm wide silicon fins, and results in increased threshold voltage. In practical terms, this means that a MOSFET with such nanoscale silicon fins, when used in digital applications, will need a higher gate (control) voltage to switch the transistor on.

Welding Nanowires

To incorporate nanowire technology into industrial applications, researchers in 2008 developed a method of welding nanowires together: a sacrificial metal nanowire is placed adjacent to the ends of the pieces to be joined (using the manipulators of a scanning electron microscope); then an electric current is applied, which fuses the wire ends. The technique fuses wires as small as 10 nm.

For nanowires with diameters less than 10 nm, existing welding techniques, which require precise control of the heating mechanism and which may introduce the possibility of damage, will not be practical. Recently scientists discovered that single-crystalline ultrathin gold nanowires with diameters ~3–10 nm can be "cold-welded" together within seconds by mechanical contact alone, and under remarkably low applied pressures (unlike macro- and micro-scale cold welding process). High-resolution transmission electron microscopy and in situ measurements reveal that the welds are nearly perfect, with the same crystal orientation, strength and electrical conductivity as the rest of the nanowire. The high quality of the welds is attributed to the nanoscale sample dimensions, oriented-attachment mechanisms and mechanically assisted fast surface diffusion. Nanowire welds were also demonstrated between gold and silver, and silver nanowires (with diameters ~5–15 nm) at near room temperature, indicating that this technique may be generally applicable for ultrathin metallic nanowires. Combined with other nano- and microfabrication technologies, cold welding is anticipated to have potential applications in the future bottom-up assembly of metallic one-dimensional nanostructures.

Applications of Nanowires

Electronic Devices

Nanowires can be used for transistors. Transistors are used widely as fundamental building element in today's electronic circuits. As predicted by Moore's law, the dimension of transistors is shrinking smaller and smaller into nanoscale. One of the key challenges of building future nanoscale transistors is ensuring good gate control over the channel. Due to the high aspect ratio, if the gate dielectric is wrapped around the nanowire channel, we can get good control of channel electrostatic potential, thereby turning the transistor on and off efficiently.

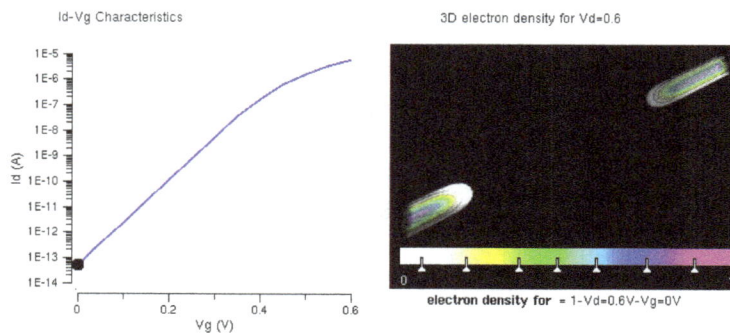

Atomistic simulation result for formation of inversion channel (electron density) and attainment of threshold voltage (IV) in a nanowire MOSFET. Note that the threshold voltage for this device lies around 0.45V.

To create active electronic elements, the first key step was to chemically dope a semiconductor nanowire. This has already been done to individual nanowires to create p-type and n-type semiconductors.

The next step was to find a way to create a p–n junction, one of the simplest electronic devices. This was achieved in two ways. The first way was to physically cross a p-type wire over an n-type wire. The second method involved chemically doping a single wire with different dopants along the length. This method created a p-n junction with only one wire.

After p-n junctions were built with nanowires, the next logical step was to build logic gates. By connecting several p-n junctions together, researchers have been able to create the basis of all logic circuits: the AND, OR, and NOT gates have all been built from semiconductor nanowire crossings.

In August 2012, researchers reported constructing the first NAND gate from undoped silicon nanowires. This avoids the problem of how to achieve precision doping of complementary nanocircuits, which is unsolved. They were able to control the Schottky barrier to achieve low-resistance contacts by placing a silicide layer in the metal-silicon interface.

It is possible that semiconductor nanowire crossings will be important to the future of digital computing. Though there are other uses for nanowires beyond these, the only ones that actually take advantage of physics in the nanometer regime are electronic.

In addition, nanowires are also being studied for use as photon ballistic waveguides as interconnects in quantum dot/quantum effect well photon logic arrays. Photons travel inside the tube, electrons travel on the outside shell.

When two nanowires acting as photon waveguides cross each other the juncture acts as a quantum dot.

Conducting nanowires offer the possibility of connecting molecular-scale entities in a molecular computer. Dispersions of conducting nanowires in different polymers are being investigated for use as transparent electrodes for flexible flat-screen displays.

Because of their high Young's moduli, their use in mechanically enhancing composites is being investigated. Because nanowires appear in bundles, they may be used as tribological additives to improve friction characteristics and reliability of electronic transducers and actuators.

Because of their high aspect ratio, nanowires are also uniquely suited to dielectrophoretic manipulation, which offers a low-cost, bottom-up approach to integrating suspended dielectric metal oxide nanowires in electronic devices such as UV, water vapor, and ethanol sensors.

Sensing of Proteins and Chemicals Using Semiconductor Nanowires

In an analogous way to FET devices in which the modulation of conductance (flow of electrons/holes) in the semiconductor, between the input (source) and the output (drain) terminals, is controlled by electrostatic potential variation (gate-electrode) of the charge carriers in the device conduction channel, the methodology of a Bio/Chem-FET is based on the detection of the local change in charge density, or so-called "field effect", that characterizes the recognition event between a target molecule and the surface receptor.

This change in the surface potential influences the Chem-FET device exactly as a 'gate' voltage does, leading to a detectable and measurable change in the device conduction. When these devices are fabricated using semiconductor nanowires as the transistor element the binding of a chemical or biological species to the surface of the sensor can lead to the depletion or accumulation of charge carriers in the "bulk" of the nanometer diameter nanowire i.e. (small cross section available for conduction channels). Moreover, the wire, which serves as a tunable conducting channel, is in close contact with the sensing environment of the target, leading to a short response time, along with orders of magnitude increase in the sensitivity of the device as a result of the huge S/V ratio of the nanowires.

While several inorganic semiconducting materials such as Si, Ge, or metal oxides (e.g. In_2O_3, SnO_2, ZnO, etc.) have been used for the preparation of nanowires. Silicon nanowires are usually the material of choice when fabricating nanowire FET-based chemo/biosensors.

Several examples of the use of silicon nanowire sensing devices include the ultra sensitive, real-time sensing of biomarker proteins for cancer, detection of single virus particles, and the detection of nitro-aromatic explosive materials such as 2,4,6 Tri-nitrotoluene (TNT) in sensitives superior to these of canines. Silicon nanowires could also be used in their twisted form, as electromechanical devices, to measure intermolecular forces with great precision.

Limitations of Sensing With Silicon Nanowire FET Devices

Generally, the charges on dissolved molecules and macromolecules are screened by dissolved counterions, since in most cases molecules bound to the devices are separated from the sensor surface by approximately 2–12 nm (the size of the receptor proteins or DNA linkers bound to the sensor surface). As a result of the screening, the electrostatic potential that arises from charges on the analyte molecule decays exponentially toward zero with distance. Thus, for optimal sensing, the Debye length must be carefully selected for nanowire FET measurements. One approach of overcoming this limitation employs fragmentation of the antibody-capturing units and control over surface receptor density, allowing more intimate binding to the nanowire of the target protein. This approach proved useful for dramatically enhancing the sensitivity of cardiac biomarkers (e.g. Troponin) detection directly from serum for the diagnosis of acute myocardial infarction.

Nanocellulose

Nanocellulose is a term referring to nano-structured cellulose. This may be either cellulose nanofibers (CNF) also called microfibrillated cellulose (MFC), nanocrystalline cellulose (NCC), or bacterial nanocellulose, which refers to nano-structured cellulose produced by bacteria.

CNF is a material composed of nanosized cellulose fibrils with a high aspect ratio (length to width ratio). Typical lateral dimensions are 5–20 nanometers and longitudinal dimension is in a wide range, typically several micrometers. It is pseudo-plastic and exhibits the property of certain gels or fluids that are thick (viscous) under normal conditions, but flow (become thin, less viscous) over time when shaken, agitated, or otherwise stressed. This property is known as thixotropy. When the shearing forces are removed the gel regains much of its original state. The fibrils are isolated from any cellulose containing source including wood-based fibers (pulp fibers) through high-pressure, high temperature and high velocity impact homogenization, grinding or microfluidization.

Nanocellulose

Nanocellulose can also be obtained from native fibers by an acid hydrolysis, giving rise to highly crystalline and rigid nanoparticles (often referred to as CNC or nanowhiskers) which are shorter (100s to 1000 nanometers) than the nanofibrils obtained through the homogenization, microfluiodization or grinding routes. The resulting material is known as nanocrystalline cellulose (NCC).

History and Terminology

The terminology microfibrillated/nanocellulose or (MFC) was first used by Turbak, Snyder and Sandberg in the late 1970s at the ITT Rayonier labs in Whippany, New Jersey, USA to describe a product prepared as a gel type material by passing wood pulp through a Gaulin type milk homogenizer at high temperatures and high pressures followed by ejection impact against a hard surface. The terminology (MFC) first appeared publicly in the early 1980s when a number of patents and publications were issued to ITT Rayonier on this totally new nanocellulose composition of matter. In later work Herrick at Rayonier also published work on making a dry powder form of the gel. Since Rayonier is one of the world's premier producers of purified pulps their business interests have always been 1) to create new uses and new markets for pulps and 2) never to compete with new or potentially new customers. Thus, as the patents issued, Rayonier gave free license to whoever wanted to pursue this new use for cellulose. Rayonier,as a company, never pursued scale-up. Rather, Turbak et al. pursued 1) finding new uses for the MFC/nanocellulose. These included using MFC as a thickener and binder in foods, cosmetics, paper formation, textiles, nonwovens, etc. and 2) evaluate swelling and other techniques for lowering the energy requirements for MFC/Nanocellulose production. ITT closed the Rayonier Whippany Labs in 1983–84 and further work on making a dry powder form of MFC was done by Herric at the Rayonier labs in Shelton, Washington, USA

The field was later taken up in Japan in the mid 1990s by the group of Taniguchi and co-workers and later by Yano and co-workers. and a host of major companies (see numerous U.S. patents issued to P&G, J&J, 3M, McNeil, etc. using U.S. patent search under inventor name Turbak search base). Today, there are still extensive research and development efforts around the world in this field.

As of August 2012, NCC was produced in a factory operated by CelluForce, producing one ton per day, and by a just-opened facility operated by the US Forest Service. Recent developments (as of January 2016) show a renewed effort to industrialise MFC and find new applications fields with newly founded companies such as Borregaard's Exilva, FiberLean (a joint-venture of Imerys and Omya), American Process and WEIDMANN Fiber Technology among others.

Manufacture

Nanocellulose/CNF or NCC can be prepared from any cellulose source material, but woodpulp is normally used.

The nanocellulose fibrils may be isolated from the wood-based fibers using mechanical methods which expose the pulp to high shear forces, ripping the larger wood-fibres apart into nanofibers. For this purpose high-pressure homogenizers, ultrasonic homogenizers, grinders or microfluidizers can be used. The homogenizers are used to delaminate the cell walls of the fibers and liberate the nanosized fibrils. This process is responsible for the high energy consumptions associated with the fiber delamination. Values over 30 MWh/tonne are not uncommon.

Pre-treatments are sometimes used to address this problem. Examples of such pre-treatments are enzymatic/mechanical pre-treatment and introduction of charged groups e.g. through carboxymethylation or TEMPO-mediated oxidation. It has been shown that energy consumption can be heavily decreased using these pre-treatments and Lindström and Ankerfors has reported values below 1 MWh/tonne.

Cellulose nanowhiskers, a more crystalline form of nanocellulose, are formed by the acid hydrolysis of native cellulose fibers commonly using sulfuric or hydrochloric acid. The amorphous sections of native cellulose are hydrolysed and after careful timing, the crystalline sections can be retrieved from the acid solution by centrifugation and washing. Cellulose nanowhiskers are rodlike highly crystalline particles (relative crystallinity index above 75%) with a rectangular cross section. Their dimensions depend on the native cellulose source material, and hydrolysis time and temperature.

Recent breakthroughs in nanocellulose production were announced April 2013 by R. Malcolm Brown Jr. PHD, a biology professor at the University of Texas at Austin, who presented his team's findings at an American Chemical Society conference in New Orleans.

Borregaard AS, a Norwegian biorefinery company based in Sarpsborg, Norway, announced in October 2014 that they are investing in a facility to produce microfibrillated cellulose on a commercial scale. The production facility will have a capacity of 1000 tonnes per year, and is scheduled to be operational from Q3, 2016.

American Process Inc, a biorefinery technology development company headquartered in Atlanta, Georgia, has also recently announced a manufacturing process using sulfur dioxide and ethanol pretreatment of biomass to produce both cellulose nanofibrils and cellulose nanocrystals and lignin-coated, hydrophobic varieties of each.

ICAR-CIRCOT Nanocellulose Pilot Plant, a unique facility for production of nanocellulose from cotton and other agro-biomass, was inaugurated by Padma Visbhushan Dr. R. A. Mashelkar in Mumbai, India on August 21, 2015. This pilot plant was funded by World Bank supported National Agricultural Innovation Project of Indian Council of Agricultural Research, New Delhi. This pilot plant has the capacity of 10 kg nanocellulose production per shift of 8 hour operation and uses patented energy efficient and eco-friendly technologies. Both nanocrystalline cellulose and nanofibrillated cellulose can be produced in this plant (both in aqueous and dry form). This is established to act as an incubation center for new entrepreneurs and as technology demonstration unit.

Structure and Properties

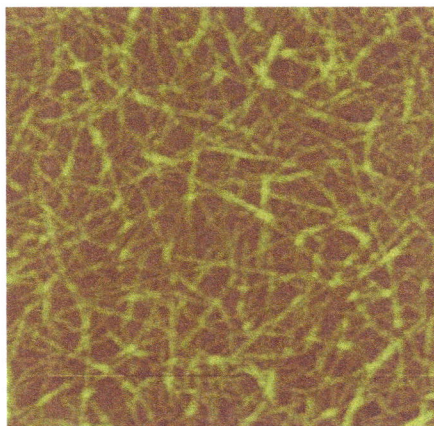

AFM height image of carboxymethylated nanocellulose adsorbed on a silica surface. The scanned surface area is 1μm2.

Nanocellulose Dimensions and Crystallinity

The ultrastructure of cellulose derived from various sources has been extensively studied. Techniques such as transmission electron microscopy (TEM), scanning electron microscopy (SEM), atomic force microscopy (AFM), wide angle X-ray scattering (WAXS), small incidence angle X-ray diffraction and solid state ^{13}C cross-polarization magic angle spinning (CP/MAS) nuclear magnetic resonance (NMR) spectroscopy have been used to characterize nanocellulose morphology. These methods have typically been applied for the investigation of dried nanocellulose morphology.

Although a combination of microscopic techniques with image analysis can provide information on nanocellulose fibril widths, it is more difficult to determine nanocellulose fibril lengths because of entanglements and difficulties in identifying both ends of individual nanofibrils. It has been reported that nanocellulose suspensions may not be homogeneous and that they consist of various structural components, including cellulose nanofibrils and nanofibril bundles.

Most methods have typically been applied to investigation of dried nanocellulose dimensions, although a study was conducted where the size and size-distribution of enzymatically pre-treated nanocellulose fibrils in a suspension was studied using cryo-TEM. The fibrils were found to be rather mono-dispersed mostly with a diameter of ca. 5 nm although occasionally thicker fibril bundles were present. It should be noted that, some newly published results indicated that by combining ultrasonication with an "oxidation pretreatment", cellulose microfibrils with a lateral dimension that belows 1 nm is observed by AFM. The lower end of the thickness dimension is around 0.4 nm, which is related to the thickness of a cellulose monolayer sheet.

The aggregate widths can be determined by CP/MAS NMR developed by Innventia AB, Sweden, which also has been demonstrated to work for nanocellulose (enzymatic pre-treatment). An average width of 17 nm has been measured with the NMR-method, which corresponds well with SEM and TEM. Using TEM, values of 15 nm have been reported for nanocellulose from carboxymethylated pulp. However, also thinner fibrils can also be detected. Wågberg et al. reported fibril widths of 5–15 nm for a nanocellulose with a charge density of about 0.5 meq./g. The group of Isogai reported fibril widths of 3–5 nm for TEMPO-oxidized cellulose having a charge density of 1.5 meq./g.

The influence of cellulose pulp chemistry on the nanocellulose microstructure has been investigated using AFM to compare the microstructure of two types of nanocellulose prepared at Innventia AB (enzymatically pre-treated nanocellulose and carboxymethylated nanocellulose). Due to the chemistry involved in producing carboxymethylated nanocellulose, it differs significantly from the enzymatically pre-treated one. The number of charged groups on the fibril surfaces are very different. The carboxymethylation pre-treatment makes the fibrils highly charged and, hence, easier to liberate, which results in smaller and more uniform fibril widths (5–15 nm) compared to the enzymatically pre-treated nanocellulose, where the fibril widths were 10–30 nm. The degree of crystallinity and the cellulose crystal structure of nanocellulose were also studied at the same time. The results clearly showed the nanocellulose exhibited cellulose crystal I organization and that the degree of crystallinity was unchanged by the preparation of the nanocellulose. Typical values for the degree of crystallinity were around 63%.

Viscosity

The unique rheology of nanocellulose dispersions was recognized by the early investigators. The high viscosity at low nanocellulose concentrations makes nanocellulose very interesting as a non-caloric stabilizer and gellant in food applications, the major field explored by the early investigators.

The dynamic rheological properties were investigated in great detail and revealed that the storage and loss modulus were independent of the angular frequency at all nanocellulose concentrations between 0.125% to 5.9%. The storage modulus values are particularly high (104 Pa at 3% concentration) compared to results for cellulose nanowhiskers (102 Pa at 3% concentration). There is also a particular strong concentration dependence as the storage modulus increases 5 orders of magnitude if the concentration is increased from 0.125% to 5.9%.

Nanocellulose gels are also highly shear thinning (the viscosity is lost upon introduction of the shear forces). The shear-thinning behaviour is particularly useful in a range of different coating applications.

Mechanical Properties

Crystalline cellulose has interesting mechanical properties for use in material applications. Its tensile strength is about 500MPa, similar to that of aluminium. Its stiffness is about 140–220 GPa, comparable with that of Kevlar and better than that of glass fiber, both of which are used commercially to reinforce plastics. Films made from nanocellulose have high strength (over 200 MPa), high stiffness (around 20 GPa) and high strain (12%). Its strength/weight ratio is 8 times that of stainless steel.

Barrier Properties

In semi-crystalline polymers, the crystalline regions are considered to be gas impermeable. Due to relatively high crystallinity, in combination with the ability of the nanofibers to form a dense network held together by strong inter-fibrillar bonds (high cohesive energy density), it has been suggested that nanocellulose might act as a barrier material. Although the number of reported oxygen permeability values are limited, reports attribute high oxygen barrier properties to nanocellulose films. One study reported an oxygen permeability of 0.0006 $(cm^3 \, \mu m)/(m^2 \, day \, kPa)$ for a ca. 5 μm thin nanocellulose film at 23 °C and 0% RH. In a related study, a more than 700-fold decrease in oxygen permeability of a polylactide (PLA) film when a nanocellulose layer was added to the PLA surface was reported.

The influence of nanocellulose film density and porosity on film oxygen permeability has recently been explored. Some authors have reported significant porosity in nanocellulose films, which seems to be in contradiction with high oxygen barrier properties, whereas Aulin et al. measured a nanocellulose film density close to density of crystalline cellulose (cellulose Iß crystal structure, 1.63 g/cm^3) indicating a very dense film with a porosity close to zero.

Changing the surface functionality of the cellulose nanoparticle can also affect the permeability of nanocellulose films. Films constituted of negatively charged cellulose nanowhiskers could effectively reduce permeation of negatively charged ions, while leaving neutral ions virtually unaffected. Positively charged ions were found to accumulate in the membrane.

Multi-Parametric Surface Plasmon Resonance is one of the methods to study barrier properties of natural, modified or coated nanocellulose. The different antifouling, moisture, solvent, antimicrobial barrier formulation quality can be measured on the nanoscale. The adsorption kinetics as well as the degree of swelling can be measured in real-time and label-free.

Foams

Nanocellulose can also be used to make aerogels/foams, either homogeneously or in composite formulations. Nanocellulose-based foams are being studied for packaging applications in order to replace polystyrene-based foams. Svagan et al. showed that nanocellulose has the ability to reinforce starch foams by using a freeze-drying technique. The advantage of using nanocellulose instead of wood-based pulp fibers is that the nanofibrils can reinforce the thin cells in the starch foam. Moreover, it is possible to prepare pure nanocellulose aerogels applying various freeze-drying and super critical CO_2 drying techniques. Aerogels and foams can be used as porous templates. Tough ultra-high porosity foams prepared from cellulose I nanofibril suspensions were studied by Sehaquiet al. A wide range of mechanical properties including compression was obtained by controlling density and nanofibril interaction in the foams. Cellulose nanowhiskers could also be made to gel in water under low power sonication giving rise to aerogels with the highest reported surface area (>600m2/g) and lowest shrinkage during drying (6.5%) of cellulose aerogels. In another study by Aulin et al., the formation of structured porous aerogels of nanocellulose by freeze-drying was demonstrated. The density and surface texture of the aerogels was tuned by selecting the concentration of the nanocellulose dispersions before freeze-drying. Chemical vapour deposition of a fluorinated silane was used to uniformly coat the aerogel to tune their wetting properties towards non-polar liquids/oils. The authors demonstrated that it is possible to switch the wettability behaviour of the cellulose surfaces between super-wetting and super-repellent, using different scales of roughness and porosity created by the freeze-drying technique and change of concentration of the nanocellulose dispersion. Structured porous cellulose foams can however also be obtained by utilizing the freeze-drying technique on cellulose generated by Gluconobacter strains of bacteria, which bio-synthesize open porous networks of cellulose fibers with relatively large amounts of nanofibrills dispersed inside. Olsson et al. demonstrated that these networks can be further impregnated with metalhydroxide/oxide precursors, which can readily be transformed into grafted magnetic nanoparticles along the cellulose nanofibers. The magnetic cellulose foam may allow for a number of novel applications of nanocellulose and the first remotely actuated magnetic super sponges absorbing 1 gram of water within a 60 mg cellulose aerogel foam were reported. Notably, these highly porous foams (>98% air) can be compressed into strong magnetic nanopapers, which may find use as functional membranes in various applications.

Surface Modification

The surface modification of nanocellulose is currently receiving a large amount of attention. Nanocellulose displays a high concentration of hydroxyl groups at the surface which can be reacted. However, hydrogen bonding strongly affects the reactivity of the surface hydroxyl groups. In addition, impurities at the surface of nanocellulose such as glucosidic and lignin fragments need to be removed before surface modification to obtain acceptable reproducibility between different batches.

Cellulose nanofibers can also be modified as cationic. This cationic cellulose increase the affinity for anions

Safety Aspects

Health, safety and environmental aspects of nanocellulose have been recently evaluated. Processing of nanocellulose does not cause significant exposure to fine particles during friction grinding or spray drying. No evidence of inflammatory effects or cytotoxicity on mouse or human macrophages can be observed after exposure to nanocellulose. The results of toxicity studies suggest that nanocellulose is not cytotoxic and does not cause any effects on inflammatory system in macrophages. In addition, nanocellulose is not acutely toxic to Vibrio fischeri in environmentally relevant concentrations.

Applications

The properties of nanocellulose (e.g. mechanical properties, film-forming properties, viscosity etc.) makes it an interesting material for many applications and the potential for a multibillion-dollar industry.

Paper and Paperboard

There is potential of nanocellulose applications in the area of paper and paperboard manufacture. Nanocelluloses are expected to enhance the fiber-fiber bond strength and, hence, have a strong reinforcement effect on paper materials. Nanocellulose may be useful as a barrier in grease-proof type of papers and as a wet-end additive to enhance retention, dry and wet strength in commodity type of paper and board products.

Composite

As described above the properties of the nanocellulose makes an interesting material for reinforcing plastics. Nanocellulose has been reported to improve the mechanical properties of, for example, thermosetting resins, starch-based matrixes, soy protein, rubber latex, poly(lactide). The composite applications may be for use as coatings and films, paints, foams, packaging.

Food

Nanocellulose can be used as a low calorie replacement for today's carbohydrate additives used as thickeners, flavour carriers and suspension stabilizers in a wide variety of food products and is useful for producing fillings, crushes, chips, wafers, soups, gravies, puddings etc. The food applications were early recognised as a highly interesting application field for nanocellulose due to the rheological behaviour of the nanocellulose gel.

Hygiene and Absorbent Products

Different applications in this field include:

- Super water absorbent (e.g. for incontinence pads material)
- Nanocellulose used together with super absorbent polymers
- Use of nanocellulose in tissue, non-woven products or absorbent structures
- Use as antimicrobial films

Emulsion and Dispersion

Apart from the numerous applications in the area of food additives, the general area of emulsion and dispersion applications in other fields has also received some attention. Oil in water applications were early recognized. The area of non-settling suspensions for pumping sand, coal as well as paints and drilling muds was also explored by the early investigators.

Oil Recovery

Hydrocarbon fracturing of oil-bearing formations is a potentially interesting and large-scale application. Nanocellulose has been suggested for use in oil recovery applications as a fracturing fluid. Drilling muds based on nanocellulose have also been suggested.

Medical, Cosmetic and Pharmaceutical

The use of nanocellulose in cosmetics and pharmaceuticals was also early recognized. A wide range of high-end applications have been suggested:

- Freeze-dried nanocellulose aerogels used in sanitary napkins, tampons, diapers or as wound dressing

- The use of nanocellulose as a composite coating agent in cosmetics e.g. for hair, eyelashes, eyebrows or nails

- A dry solid nanocellulose composition in the form of tablets for treating intestinal orders

- Nanocellulose films for screening of biological compounds and nucleic acids encoding a biological compound

- Filter medium partly based on nanocellulose for leukocyte free blood transfusion

- A buccodental formulation, comprising nanocellulose and a polyhydroxylated organic compound

- Powdered nanocellulose has also been suggested as an excipient in pharmaceutical compositions

- nanocellulose in compositions of a photoreactive noxious substance purging agent

- Elastic cryo-structured gels for potential biomedical and biotechnological application.

- Matrix for 3D cell culture

Other Applications

- Activate the dissolution of cellulose in different solvents

- Regenerated cellulose products, such as fibers films, cellulose derivatives

- Tobacco filter additive

- Organometallic modified nanocellulose in battery separators

- Reinforcement of conductive materials

- Loud-speaker membranes

- High-flux membranes

- Flexible electronic displays

- Computer components

- Capacitors

- Lightweight body armour and ballistic glass

- Corrosion inhibitors

References

- Godovsky, D. Y. (2000). "Device Applications of Polymer-Nanocomposites". In Chang, J. Y. Biopolymers · PVA Hydrogels, Anionic Polymerisation Nanocomposites. Advances in Polymer Science. 153. pp. 163–205. doi:10.1007/3-540-46414-X_4. ISBN 978-3-540-67313-2.

- Kolosnjaj J, Szwarc H, Moussa F (2007). "Toxicity studies of carbon nanotubes". Adv Exp Med Biol. Advances in Experimental Medicine and Biology. 620: 181–204. doi:10.1007/978-0-387-76713-0_14. ISBN 978-0-387-76712-3. PMID 18217344.

- Pacios Pujadó, Mercè (2012). Carbon Nanotubes as Platforms for Biosensors with Electrochemical and Electronic Transduction. Springer Heidelberg. pp. XX,208. doi:10.1007/978-3-642-31421-6. ISBN 978-3-642-31421-6. ISSN 2190-5053.

- Turbak, A.F.; F.W. Snyder; K.R. Sandberg (1983). "Microfibrillated cellulose, a new cellulose product: Properties, uses and commercial potential". In A. Sarko (ed.). Proceedings of the Ninth Cellulose Conference. Applied Polymer Symposia, 37. New York City: Wiley. pp. 815–827. ISBN 0-471-88132-5.

- Herrick, F.W.; R.L. Casebier; J.K. Hamilton; K.R. Sandberg (1983). "Microfibrillated cellulose: morphology and accessibility". In A. Sarko (ed.). Proceedings of the Ninth Cellulose Conference. Applied Polymer Symposia, 37. New York City: Wiley. pp. 797–813. ISBN 0-471-88132-5.

- Berglund, Lars (2005). "Cellulose-based nanocomposites". In A.K. Mohanty; M. Misra; L. Drzal (Eds). Natural fibers, biopolymers and biocomposites. Boca Raton, Florida: CRC Press. pp. 807–832. ISBN 978-0-8493-1741-5.

- Lindström, Tom; Mikael Ankerfors (2009). "NanoCellulose Developments in Scandinavia". 7th International Paper and Coating Chemistry Symposium (Preprint CD ed.). Hamilton, Ontario: McMaster University Engineering. ISBN 978-0-9812879-0-4.

- Chinga-Carrasco, G., Miettinen, A., Luengo Hendriks, C.-L., Gamstedt, E.K., Kataja, M. (2011). Structural Characterisation of Kraft Pulp Fibres and Their Nanofibrillated Materials for Biodegradable Composite Applications. InTech. ISBN 978-953-307-352-1.

- "Green sonochemical synthesis of silver nanoparticles at varying concentrations of κ-carrageenan" (PDF). springer.com. Retrieved 2016-02-15.

4

Technologies Related to Nanocomposites

Nanocomposites find application in various fields and there are several technologies that singularly depend on nanocomposites. This chapter lists and describes technologies like titanate nanosheet, ceramic engineering and polyethylene terephthalate. The topics discussed in the chapter are of great importance to broaden the existing knowledge on nanocomposites.

Titanate Nanosheet

Titanate (IV) nanosheets (TiNSs) have a 2D structure where TiO_6 octahedra are edge-linked in a lepidocrocite-type 2D lattice with chemical formula $H_xTi_2 - _{x/4}$ $_{x/4}O_4$ H_2O (x~0.7; , vacancy). Titanate nanosheets may be regarded as sheets with molecular thickness and infinite planar dimensions. TiNSs are typically formed via liquid-phase exfoliation of protonic titanate. In inorganic layered materials, individual layers are bound to each other by van der Waals interactions if they are neutral, and additional Coulomb interactions if they are composed of oppositely charged layers. Through liquid-phase exfoliation, these individual sheets of layered materials can be efficiently separated using an appropriate solvent, creating single-layer colloidal suspensions. Solvents must have an interaction energy with the layers that is greater than the interaction energy between two layers. *In situ* X-Ray diffraction data indicates that TiNSs can be treated as macromolecules with a sufficient amount of solvent in between layers so that they behave as individual sheets.

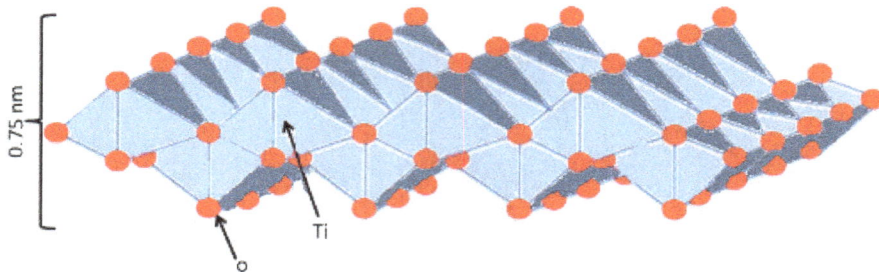

Structure of titanate nanosheets

Properties

Unilamellar TiNSs have a number of unique properties, and are said to combine those of conventional titanate and titania. Structurally, they are infinite ultrathin (~0.75 nm) 2D sheets with a high density of negative surface charges originating from the oxygen atoms at the corners of the adjoint octahedrons . TiNSs may balance this anionic charge by inserting a counterionic layer between the two sheets either via layering or in aqueous solution. This electric double layer gives the material flexible interlayer distances, high cation exchange capacity, and excellent dielectric capabilities.

Typically, titanium oxide suffers from oxygen vacancies, which diminish its potential as capacitors due to vacancies acting as high-leakage paths and charge carrier traps, however, TiNSs possess Ti vacancies, which promote channels for electron transfer. When Ti vacancies are present, the effective charge felt by electrons on oxygen atoms reduces and allows less hindered electron motion.

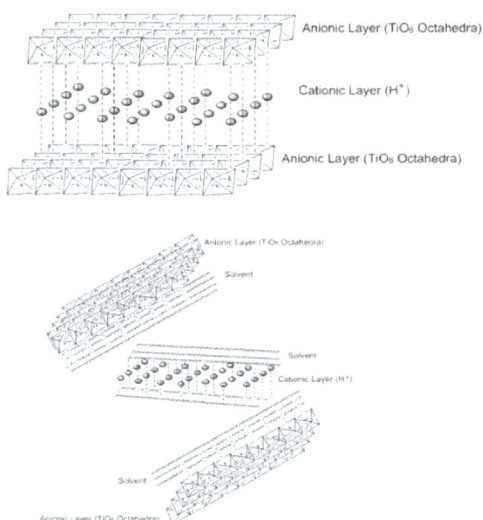

A starting inorganic material composed of alternating layers of charged material, in the case of protonated titanate the cationic layer is composed of protons while the anionic layer is composed of edge linked TiO_6 octahedra. A solvent is chosen such that it has a greater interaction energy with the sheets than they do with each other, and this interaction will replace the bonds holding the sheets together yielding colloidal suspensions of 2D nanosheets.

Applications

TiNSs can act as highly efficient adsorbents and photocatalysts due to their two-dimensional geometry and structure. This phenomenon can be exploited for several applications, including the removal of metal ions and dyes from water systems. Further, TiNSs potential as an electrocatalyst may enhance fuel cell efficiency during fuel oxidation. Similarly, intercalated myoglobin is proven to be an efficient catalyst for H_2O_2.

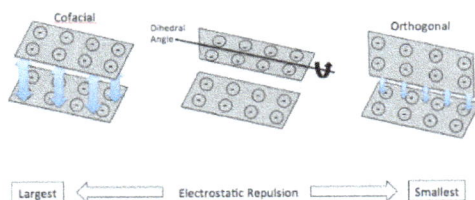

Cofacial Alignment of TiNSs. The cofacial alignment of the anionically charged titanate maximizes the repulsion between cofacial sheets and occurs under a magnetic field.

TiNSs may also be used for immobilizing biomolecules. When a monolayer of hemoglobin is intercalated into TiNSs, the electron transfer between the active sites of the protein and the electrodes is amplified, and electrocatalytic activity for O_2 reduction increases. In addition, heterostructured nanosheets of Fe_3O_4-$Na_2Ti_3O_7$ can be used for protein separation. When placed in the aqueous environment at pH 6, positively charged hemoglobin binds to the nanosheets, whereas negative albumin can be detected in the supernatant.

Perhaps the most interesting application of TiNSs is in the development of a material dominated by electrostatically repulsive interactions. TiNSs exhibit maximum electrostatic repulsion when they are aligned cofacially. To create the hydrogel based on this, a solution of TiNSs is placed in a strong magnetic field where repulsive forces induce a quasi-crystalline structure. Upon irradiation with UV-light, the solution polymerizes and creates cross-linked network, which is non-covalently attached to the TiNSs. This creates a composite that resists orthogonally applied compressive forces, but easily deforms due to shear forces. TiNSs solutions of this sort may be used as an anti-vibration or vibration-isolating material and in the design of artificial cartilage.

Ceramic Engineering

Ceramic engineering is the science and technology of creating objects from inorganic, non-metallic materials. This is done either by the action of heat, or at lower temperatures using precipitation reactions from high-purity chemical solutions. The term includes the purification of raw materials, the study and production of the chemical compounds concerned, their formation into components and the study of their structure, composition and properties.

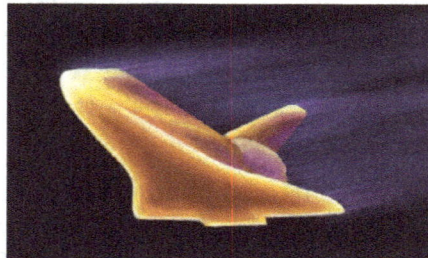

Simulation of the outside of the Space Shuttle as it heats up to over 1,500 °C (2,730 °F) during re-entry into the Earth's atmosphere

Ceramic materials may have a crystalline or partly crystalline structure, with long-range order on atomic scale. Glass ceramics may have an amorphous or glassy structure, with limited or short-range atomic order. They are either formed from a molten mass that solidifies on cooling, formed and matured by the action of heat, or chemically synthesized at low temperatures using, for example, hydrothermal or sol-gel synthesis.

Bearing components made from 100% silicon nitride Si_3N_4

The special character of ceramic materials gives rise to many applications in materials engineering, electrical engineering, chemical engineering and mechanical engineering. As ceramics are heat resistant, they can be used for many tasks for which materials like metal and polymers are unsuitable. Ceramic materials are used in a wide range of industries, including mining, aerospace, medicine, refinery, food and chemical industries, packaging science, electronics, industrial and transmission electricity, and guided lightwave transmission.

Ceramic bread knife

History

The word "ceramic" is derived from the Greek word (*keramikos*) meaning pottery. It is related to the older Indo-European language root "to burn", "Ceramic" may be used as a noun in the singular to refer to a ceramic material or the product of ceramic manufacture, or as an adjective. The plural "ceramics" may be used to refer the making of things out of ceramic materials. Ceramic engineering, like many sciences, evolved from a different discipline by today's standards. Materials science engineering is grouped with ceramics engineering to this day.

Leo Morandi's tile glazing line (circa 1945)

Abraham Darby first used coke in 1709 in Shropshire, England, to improve the yield of a smelting process. Coke is now widely used to produce carbide ceramics. Potter Josiah Wedgwood opened the first modern ceramics factory in Stoke-on-Trent, England, in 1759. Austrian chemist Carl Josef Bayer, working for the textile industry in Russia, developed a process to separate alumina from bauxite ore in 1888. The Bayer process is still used to purify alumina for the ceramic and aluminium industries. Brothers Pierre and Jacques Curie discovered piezoelectricity in Rochelle salt circa 1880. Piezoelectricity is one of the key properties of electroceramics.

E.G. Acheson heated a mixture of coke and clay in 1893, and invented carborundum, or synthetic silicon carbide. Henri Moissan also synthesized SiC and tungsten carbide in his electric arc furnace in Paris about the same time as Acheson. Karl Schröter used liquid-phase sintering to bond or "cement" Moissan's tungsten carbide particles with cobalt in 1923 in Germany. Cemented (metal-bonded) carbide edges greatly increase the durability of hardened steel cutting tools. W.H. Nernst developed cubic-stabilized zirconia in the 1920s in Berlin. This material is used as an oxygen sensor in exhaust systems. The main limitation on the use of ceramics in engineering is brittleness.

Military

The military requirements of World War II encouraged developments, which created a need for high-performance materials and helped speed the development of ceramic science and engineering. Throughout the 1960s and 1970s, new types of ceramics were developed in response to advances in atomic energy, electronics, communications, and space travel. The discovery of ceramic superconductors in 1986 has spurred intense research to develop superconducting ceramic parts for electronic devices, electric motors, and transportation equipment.

Soldiers pictured during the 2003 Iraq War seen through IR transparent Night Vision Goggles

There is an increasing need in the military sector for high-strength, robust materials which have the capability to transmit light around the visible (0.4–0.7 micrometers) and mid-infrared (1–5 micrometers) regions of the spectrum. These materials are needed for applications requiring transparent armour. Transparent armor is a material or system of materials designed to be optically transparent, yet protect from fragmentation or ballistic impacts. The primary requirement for a transparent armour system is to not only defeat the designated threat but also provide a multi-hit capability with minimized distortion of surrounding areas. Transparent armour windows must also be compatible with night vision equipment. New materials that are thinner, lightweight, and offer better ballistic performance are being sought. Such solid-state components have found widespread use for various applications in the electro-optical field including: optical fibres for guided lightwave transmission, optical switches, laser amplifiers and lenses, hosts for solid-state lasers and optical window materials for gas lasers, and infrared (IR) heat seeking devices for missile guidance systems and IR night vision.

Modern Industry

Now a multibillion-dollar a year industry, ceramic engineering and research has established itself as an important field of science. Applications continue to expand as researchers develop new kinds of ceramics to serve different purposes.

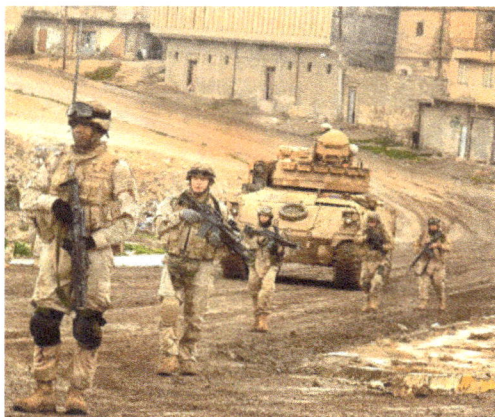

U.S. Army soldiers wearing bulletproof ballistic vests with an armoured M3 Bradley.

- Zirconium dioxide ceramics are used in the manufacture of knives. The blade of the ceramic knife will stay sharp for much longer than that of a steel knife, although it is more brittle and can be snapped by dropping it on a hard surface.

- Ceramics such as alumina, boron carbide and silicon carbide have been used in bulletproof vests to repel small arms rifle fire. Such plates are known commonly as trauma plates. Similar material is used to protect cockpits of some military aircraft, because of the low weight of the material.

- Silicon nitride parts are used in ceramic ball bearings. Their higher hardness means that they are much less susceptible to wear and can offer more than triple lifetimes. They also deform less under load meaning they have less contact with the bearing retainer walls and can roll faster. In very high speed applications, heat from friction during rolling can cause problems for metal bearings; problems which are reduced by the use of ceramics. Ceramics are also more chemically resistant and can be used in wet environments where steel bearings would rust. The major drawback to using ceramics is a significantly higher cost. In many cases their electrically insulating properties may also be valuable in bearings.

- In the early 1980s, Toyota researched production of an adiabatic ceramic engine which can run at a temperature of over 6000 °F (3300 °C). Ceramic engines do not require a cooling system and hence allow a major weight reduction and therefore greater fuel efficiency. Fuel efficiency of the engine is also higher at high temperature, as shown by Carnot's theorem. In a conventional metallic engine, much of the energy released from the fuel must be dissipated as waste heat in order to prevent a meltdown of the metallic parts. Despite all of these desirable properties, such engines are not in production because the manufacturing of ceramic parts in the requisite precision and durability is difficult. Imperfection in the ceramic leads to cracks, which can lead to potentially dangerous equipment failure. Such engines are possible in laboratory settings, but mass-production is not feasible with current technology.

- Work is being done in developing ceramic parts for gas turbine engines. Currently, even blades made of advanced metal alloys used in the engines' hot section require cooling and careful limiting of operating temperatures. Turbine engines made with ceramics could operate more efficiently, giving aircraft greater range and payload for a set amount of fuel.

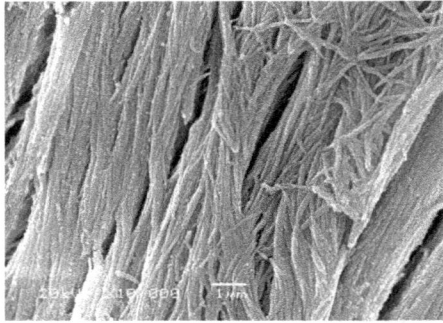

Scanning electron microscopy image of bone

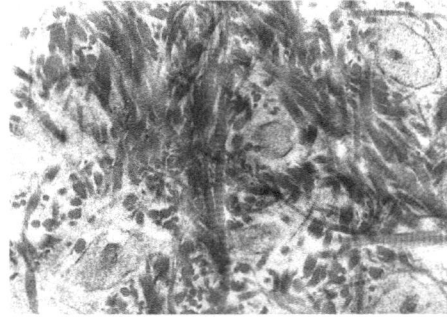

Collagen fibers of woven bone

Recently, there have been advances in ceramics which include bio-ceramics, such as dental implants and synthetic bones. Hydroxyapatite, the natural mineral component of bone, has been made synthetically from a number of biological and chemical sources and can be formed into ceramic materials. Orthopaedic implants made from these materials bond readily to bone and other tissues in the body without rejection or inflammatory reactions. Because of this, they are of great interest for gene delivery and tissue engineering scaffolds. Most hydroxyapatite ceramics are very porous and lack mechanical strength and are used to coat metal orthopaedic devices to aid in forming a bond to bone or as bone fillers. They are also used as fillers for orthopaedic plastic screws to aid in reducing the inflammation and increase absorption of these plastic materials. Work is being done to make strong, fully dense nano crystalline hydroxyapatite ceramic materials for orthopaedic weight bearing devices, replacing foreign metal and plastic orthopaedic materials with a synthetic, but naturally occurring, bone mineral. Ultimately these ceramic materials may be used as bone replacements or with the incorporation of protein collagens, synthetic bones.

- High-tech ceramic is used in watch-making for producing watch cases. The material is valued by watchmakers for its light weight, scratch-resistance, durability and smooth touch. IWC is one of the brands that initiated the use of ceramic in watch-making. The case of the IWC 2007 Top Gun edition of the Pilot's Watch Double chronograph is crafted in high-tech black ceramic.

Glass-ceramics

A high strength glass-ceramic cook-top with negligible thermal expansion.

Glass-ceramic materials share many properties with both glasses and ceramics. Glass-ceramics have an amorphous phase and one or more crystalline phases and are produced by a so-called "controlled crystallization", which is typically avoided in glass manufacturing. Glass-ceramics often contain a crystalline phase which constitutes anywhere from 30% [m/m] to 90% [m/m] of its composition by volume, yielding an array of materials with interesting thermomechanical properties.

In the processing of glass-ceramics, molten glass is cooled down gradually before reheating and annealing. In this heat treatment the glass partly crystallizes. In many cases, so-called 'nucleation agents' are added in order to regulate and control the crystallization process. Because there is usually no pressing and sintering, glass-ceramics do not contain the volume fraction of porosity typically present in sintered ceramics.

The term mainly refers to a mix of lithium and aluminosilicates which yields an array of materials with interesting thermomechanical properties. The most commercially important of these have the distinction of being impervious to thermal shock. Thus, glass-ceramics have become extremely useful for countertop cooking. The negative thermal expansion coefficient (TEC) of the crystalline ceramic phase can be balanced with the positive TEC of the glassy phase. At a certain point (~70% crystalline) the glass-ceramic has a net TEC near zero. This type of glass-ceramic exhibits excellent mechanical properties and can sustain repeated and quick temperature changes up to 1000 °C.

Processing Steps

The traditional ceramic process generally follows this sequence: Milling Batching Mixing Forming Drying Firing Assembly.

Ball mill

- Milling is the process by which materials are reduced from a large size to a smaller size. Milling may involve breaking up cemented material (in which case individual particles retain their shape) or pulverization (which involves grinding the particles themselves to a smaller size). Milling is generally done by mechanical means, including *attrition* (which is particle-to-particle collision that results in agglomerate break up or particle shearing), *compression* (which applies a forces that results in fracturing), and *impact* (which employs a milling medium or the particles themselves to cause fracturing). Attrition milling equipment includes the wet scrubber (also called the planetary mill or wet attrition mill), which has paddles in water creating vortexes in which the material collides and break up. Compression mills include the jaw crusher, roller crusher and cone crusher. Impact mills include the ball mill, which has media that tumble and fracture the material. Shaft impactors cause particle-to particle attrition and compression.

- Batching is the process of weighing the oxides according to recipes, and preparing them for mixing and drying.

- Mixing occurs after batching and is performed with various machines, such as dry mixing ribbon mixers (a type of cement mixer), Mueller mixers, and pug mills. Wet mixing generally involves the same equipment.

- Forming is making the mixed material into shapes, ranging from toilet bowls to spark plug insulators. Forming can involve: (1) Extrusion, such as extruding "slugs" to make bricks, (2) Pressing to make shaped parts, (3) Slip casting, as in making toilet bowls, wash basins and ornamentals like ceramic statues. Forming produces a "green" part, ready for drying. Green parts are soft, pliable, and over time will lose shape. Handling the green product will change its shape. For example, a green brick can be "squeezed", and after squeezing it will stay that way.

- Drying is removing the water or binder from the formed material. Spray drying is widely used to prepare powder for pressing operations. Other dryers are tunnel dryers and periodic dryers. Controlled heat is applied in this two-stage process. First, heat removes water. This step needs careful control, as rapid heating causes cracks and surface defects. The dried part is smaller than the green part, and is brittle, necessitating careful handling, since a small impact will cause crumbling and breaking.

- Sintering is where the dried parts pass through a controlled heating process, and the oxides are chemically changed to cause bonding and densification. The fired part will be smaller than the dried part.

Forming Methods

Ceramic forming techniques include throwing, slipcasting, tape casting, freeze-casting, injection moulding, dry pressing, isostatic pressing, hot isostatic pressing (HIP) and others. Methods for forming ceramic powders into complex shapes are desirable in many areas of technology. Such methods are required for producing advanced, high-temperature structural parts such as heat engine components and turbines. Materials other than ceramics which are used in these processes may include: wood, metal, water, plaster and epoxy—most of which will be eliminated upon firing.

These forming techniques are well known for providing tools and other components with dimensional stability, surface quality, high (near theoretical) density and microstructural uniformity. The increasing use and diversity of speciality forms of ceramics adds to the diversity of process technologies to be used.

Thus, reinforcing fibres and filaments are mainly made by polymer, sol-gel, or CVD processes, but melt processing also has applicability. The most widely used speciality form is layered structures, with tape casting for electronic substrates and packages being pre-eminent. Photo-lithography is of increasing interest for precise patterning of conductors and other components for such packaging. Tape casting or forming processes are also of increasing interest for other applications, ranging from open structures such as fuel cells to ceramic composites.

The other major layer structure is coating, where melt spraying is very important, but chemical and physical vapour deposition and chemical (e.g., sol-gel and polymer pyrolysis) methods are all seeing increased use. Besides open structures from formed tape, extruded structures, such as honeycomb catalyst supports, and highly porous structures, including various foams, for example, reticulated foam, are of increasing use.

Densification of consolidated powder bodies continues to be achieved predominantly by (pressureless) sintering. However, the use of pressure sintering by hot pressing is increasing, especially for non-oxides and parts of simple shapes where higher quality (mainly microstructural homogeneity) is needed, and larger size or multiple parts per pressing can be an advantage.

The Sintering Process

The principles of sintering-based methods are simple ("sinter" has roots in the English "cinder"). The firing is done at a temperature below the melting point of the ceramic. Once a roughly-held-together object called a "green body" is made, it is baked in a kiln, where atomic and molecular diffusion processes give rise to significant changes in the primary microstructural features. This includes the gradual elimination of porosity, which is typically accompanied by a net shrinkage and overall densification of the component. Thus, the pores in the object may close up, resulting in a denser product of significantly greater strength and fracture toughness.

Another major change in the body during the firing or sintering process will be the establishment of the polycrystalline nature of the solid. This change will introduce some form of grain size distribution, which will have a significant impact on the ultimate physical properties of the material. The grain sizes will either be associated with the initial particle size, or possibly the sizes of aggregates or particle clusters which arise during the initial stages of processing.

The ultimate microstructure (and thus the physical properties) of the final product will be limited by and subject to the form of the structural template or precursor which is created in the initial stages of chemical synthesis and physical forming. Hence the importance of chemical powder and polymer processing as it pertains to the synthesis of industrial ceramics, glasses and glass-ceramics.

There are numerous possible refinements of the sintering process. Some of the most common involve pressing the green body to give the densification a head start and reduce the sintering time needed. Sometimes organic binders such as polyvinyl alcohol are added to hold the green body together; these burn out during the firing (at 200–350 °C). Sometimes organic lubricants are added during pressing to increase densification. It is common to combine these, and add binders and lubricants to a powder, then press. (The formulation of these organic chemical additives is an art in itself. This is particularly important in the manufacture of high performance ceramics such as those used by the billions for electronics, in capacitors, inductors, sensors, etc.)

A slurry can be used in place of a powder, and then cast into a desired shape, dried and then sintered. Indeed, traditional pottery is done with this type of method, using a plastic mixture worked with the hands. If a mixture of different materials is used together in a ceramic, the sintering temperature is sometimes above the melting point of one minor component – a *liquid phase* sintering. This results in shorter sintering times compared to solid state sintering.

Strength of Ceramics

A material's strength is dependent on its microstructure. The engineering processes to which a material is subjected can alter this microstructure. The variety of strengthening mechanisms that alter the strength of a material include the mechanism of grain boundary strengthening. Thus, although yield strength is maximized with decreasing grain size, ultimately, very small grain sizes make the material brittle. Considered in tandem with the fact that the yield strength is the parameter that predicts plastic deformation in the material, one can make informed decisions on how to increase the strength of a material depending on its microstructural properties and the desired end effect.

The relation between yield stress and grain size is described mathematically by the Hall-Petch equation which is

where k_y is the strengthening coefficient (a constant unique to each material), σ_o is a materials constant for the starting stress for dislocation movement (or the resistance of the lattice to dislocation motion), d is the grain diameter, and σ_y is the yield stress.

Theoretically, a material could be made infinitely strong if the grains are made infinitely small. This is, unfortunately, impossible because the lower limit of grain size is a single unit cell of the material. Even then, if the grains of a material are the size of a single unit cell, then the material is in fact amorphous, not crystalline, since there is no long range order, and dislocations can not be defined in an amorphous material. It has been observed experimentally that the microstructure with the highest yield strength is a grain size of about 10 nanometres, because grains smaller than this undergo another yielding mechanism, grain boundary sliding. Producing engineering materials with this ideal grain size is difficult because of the limitations of initial particle sizes inherent to nanomaterials and nanotechnology.

Theory of Chemical Processing

Microstructural Uniformity

In the processing of fine ceramics, the irregular particle sizes and shapes in a typical powder often lead to non-uniform packing morphologies that result in packing density variations in the powder compact. Uncontrolled agglomeration of powders due to attractive van der Waals forces can also give rise to in microstructural inhomogeneities.

Differential stresses that develop as a result of non-uniform drying shrinkage are directly related to the rate at which the solvent can be removed, and thus highly dependent upon the distribution of porosity. Such stresses have been associated with a plastic-to-brittle transition in consolidated bodies, and can yield to crack propagation in the unfired body if not relieved.

In addition, any fluctuations in packing density in the compact as it is prepared for the kiln are often amplified during the sintering process, yielding inhomogeneous densification. Some pores and other structural defects associated with density variations have been shown to play a detrimental role in the sintering process by growing and thus limiting end-point densities. Differential stresses arising from inhomogeneous densification have also been shown to result in the propagation of internal cracks, thus becoming the strength-controlling flaws.

It would therefore appear desirable to process a material in such a way that it is physically uniform with regard to the distribution of components and porosity, rather than using particle size distributions which will maximize the green density. The containment of a uniformly dispersed assembly of strongly interacting particles in suspension requires total control over particle-particle interactions. Monodisperse colloids provide this potential.

Monodisperse powders of colloidal silica, for example, may therefore be stabilized sufficiently to ensure a high degree of order in the colloidal crystal or polycrystalline colloidal solid which results from aggregation. The degree of order appears to be limited by the time and space allowed for longer-range correlations to be established.

Such defective polycrystalline colloidal structures would appear to be the basic elements of sub-micrometer colloidal materials science, and, therefore, provide the first step in developing a more rigorous understanding of the mechanisms involved in microstructural evolution in inorganic systems such as polycrystalline ceramics.

Self-assembly

Self-assembly is the most common term in use in the modern scientific community to describe the spontaneous aggregation of particles (atoms, molecules, colloids, micelles, etc.) without the influence of any external forces. Large groups of such particles are known to assemble themselves into thermodynamically stable, structurally well-defined arrays, quite reminiscent of one of the 7 crystal systems found in metallurgy and mineralogy (e.g. face-centred cubic, body-centred cubic, etc.).The fundamental difference in equilibrium structure is in the spatial scale of the unit cell (or lattice parameter) in each particular case.

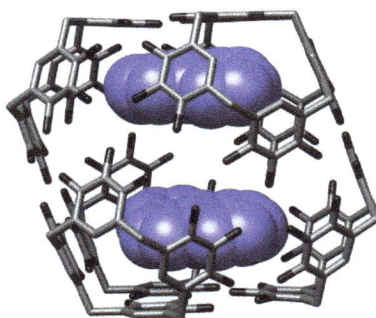

An example of a supramolecular assembly.

Thus, self-assembly is emerging as a new strategy in chemical synthesis and nanotechnology. Molecular self-assembly has been observed in various biological systems and underlies the formation of a wide variety of complex biological structures. Molecular crystals, liquid crystals, colloids, micelles, emulsions, phase-separated polymers, thin films and self-assembled monolayers all represent examples of the types of highly ordered structures which are obtained using these techniques. The distinguishing feature of these methods is self-organization in the absence of any external forces.

In addition, the principal mechanical characteristics and structures of biological ceramics, polymer composites, elastomers, and cellular materials are being re-evaluated, with an emphasis on bioinspired materials and structures. Traditional approaches focus on design methods of biological materials using conventional synthetic materials. This includes an emerging class of mechanically superior biomaterials based on microstructural features and designs found in nature. The new horizons have been identified in the synthesis of bioinspired materials through processes that are characteristic of biological systems in nature. This includes the nanoscale self-assembly of the components and the development of hierarchical structures.

Ceramic Composites

Substantial interest has arisen in recent years in fabricating ceramic composites. While there is considerable interest in composites with one or more non-ceramic constituents, the greatest attention is on composites in which all constituents are ceramic. These typically comprise two ceramic constituents: a continuous matrix, and a dispersed phase of ceramic particles, whiskers, or short (chopped) or continuous ceramic fibres. The challenge, as in wet chemical processing, is to obtain a uniform or homogeneous distribution of the dispersed particle or fibre phase.

The Porsche Carrera GT's carbon-ceramic (silicon carbide) composite disc brake

Consider first the processing of particulate composites. The particulate phase of greatest interest is tetragonal zirconia because of the toughening that can be achieved from the phase transformation from the metastable tetragonal to the monoclinic crystalline phase, aka transformation toughening. There is also substantial interest in dispersion of hard, non-oxide phases such as SiC, TiB, TiC, boron, carbon and especially oxide matrices like alumina and mullite. There is also interest too incorporating other ceramic particulates, especially those of highly anisotropic thermal expansion. Examples include Al_2O_3, TiO_2, graphite, and boron nitride.

Silicon carbide single crystal

In processing particulate composites, the issue is not only homogeneity of the size and spatial distribution of the dispersed and matrix phases, but also control of the matrix grain size. However, there is some built-in self-control due to inhibition of matrix grain growth by the dispersed phase. Particulate composites, though generally offer increased resistance to damage, failure, or both, are still quite sensitive to inhomogeneities of composition as well as other processing defects such as pores. Thus they need good processing to be effective.

Particulate composites have been made on a commercial basis by simply mixing powders of the two constituents. Although this approach is inherently limited in the homogeneity that can be achieved, it is the most readily adaptable for existing ceramic production technology. However, other approaches are of interest.

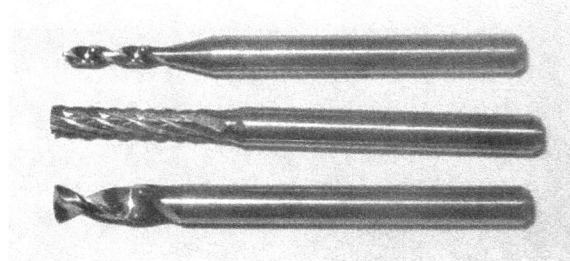

Tungsten carbide milling bits

From the technological standpoint, a particularly desirable approach to fabricating particulate composites is to coat the matrix or its precursor onto fine particles of the dispersed phase with good control of the starting dispersed particle size and the resultant matrix coating thickness. One should in principle be able to achieve the ultimate in homogeneity of distribution and thereby optimize composite performance. This can also have other ramifications, such as allowing more useful composite performance to be achieved in a body having porosity, which might be desired for other factors, such as limiting thermal conductivity.

There are also some opportunities to utilize melt processing for fabrication of ceramic, particulate, whisker and short-fibre, and continuous-fibre composites. Clearly, both particulate and whisker composites are conceivable by solid-state precipitation after solidification of the melt. This can also be obtained in some cases by sintering, as for precipitation-toughened, partially stabilized zirconia. Similarly, it is known that one can directionally solidify ceramic eutectic mixtures and hence obtain uniaxially aligned fibre composites. Such composite processing has typically been limited to very simple shapes and thus suffers from serious economic problems due to high machining costs.

Clearly, there are possibilities of using melt casting for many of these approaches. Potentially even more desirable is using melt-derived particles. In this method, quenching is done in a solid solution or in a fine eutectic structure, in which the particles are then processed by more typical ceramic powder processing methods into a useful body. There have also been preliminary attempts to use melt spraying as a means of forming composites by introducing the dispersed particulate, whisker, or fibre phase in conjunction with the melt spraying process.

Other methods besides melt infiltration to manufacture ceramic composites with long fibre reinforcement are chemical vapour infiltration and the infiltration of fibre preforms with organic precursor, which after pyrolysis yield an amorphous ceramic matrix, initially with a low density. With repeated cycles of infiltration and pyrolysis one of those types of ceramic matrix composites is produced. Chemical vapour infiltration is used to manufacture carbon/carbon and silicon carbide reinforced with carbon or silicon carbide fibres.

Besides many process improvements, the first of two major needs for fibre composites is lower fibre costs. The second major need is fibre compositions or coatings, or composite processing, to reduce degradation that results from high-temperature composite exposure under oxidizing conditions.

Applications

The products of technical ceramics include tiles used in the Space Shuttle program, gas burner nozzles, ballistic protection, nuclear fuel uranium oxide pellets, bio-medical implants, jet engine turbine blades, and missile nose cones.

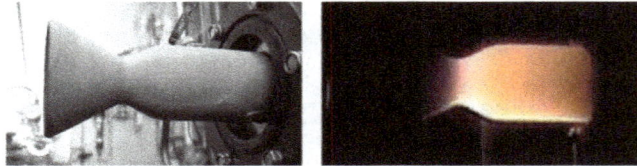

Silicon nitride thruster. Left: Mounted in test stand. Right: Being tested with H_2/O_2 propellants

Its products are often made from materials other than clay, chosen for their particular physical properties. These may be classified as follows:

- Oxides: silica, alumina, zirconia

- Non-oxides: carbides, borides, nitrides, silicides

- Composites: particulate or whisker reinforced matrices, combinations of oxides and non-oxides (e.g. polymers).

Ceramics can be used in many technological industries. One application is the ceramic tiles on NASA's Space Shuttle, used to protect it and the future supersonic space planes from the searing heat of re-entry into the Earth's atmosphere. They are also used widely in electronics and optics. In addition to the applications listed here, ceramics are also used as a coating in various engineering cases. An example would be a ceramic bearing coating over a titanium frame used for an aircraft. Recently the field has come to include the studies of single crystals or glass fibres, in addition to traditional polycrystalline materials, and the applications of these have been overlapping and changing rapidly.

Aerospace

- Engines; Shielding a hot running aircraft engine from damaging other components.

- Airframes; Used as a high-stress, high-temp and lightweight bearing and structural component.

- Missile nose-cones; Shielding the missile internals from heat.

- Space Shuttle tiles

- Space-debris ballistic shields – ceramic fiber woven shields offer better protection to hypervelocity (~7 km/s) particles than aluminium shields of equal weight.

- Rocket nozzles, withstands and focuses the exhaust of the rocket booster.

- Unmanned Air Vehicles; Implications of ceramic engine utilization in aeronautical applications (such as Unmanned Air Vehicles) may result in enhanced performance characteristics and less operational costs.

Biomedical

A titanium hip prosthesis, with a ceramic head and polyethylene acetabular cup.

- Artificial bone; Dentistry applications, teeth.
- Biodegradable splints; Reinforcing bones recovering from osteoporosis
- Implant material

Electronics

- Capacitors
- Integrated circuit packages
- Transducers
- Insulators

Optical

- Optical fibres, guided lightwave transmission
- Switches
- Laser amplifiers
- Lenses
- Infrared heat-seeking devices

Automotive

- Heat shield
- Exhaust heat management

Biomaterials

Silicification is quite common in the biological world and occurs in bacteria, single-celled organisms, plants, and animals (invertebrates and vertebrates). Crystalline minerals formed in such environment often show exceptional physical properties (e.g. strength, hardness, fracture toughness) and tend to form hierarchical structures that exhibit microstructural order over a range of length or spatial scales. The minerals are crystallized from an environment that is undersaturated with respect to silicon, and under conditions of neutral pH and low temperature (0–40 °C). Formation of the mineral may occur either within or outside of the cell wall of an organism, and specific biochemical reactions for mineral deposition exist that include lipids, proteins and carbohydrates.

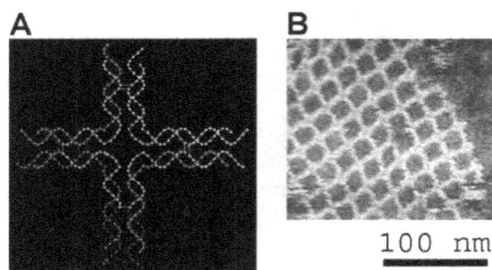

The DNA structure at left (schematic shown) will self-assemble into the structure visualized by atomic force microscopy at right.

Most natural (or biological) materials are complex composites whose mechanical properties are often outstanding, considering the weak constituents from which they are assembled. These complex structures, which have risen from hundreds of million years of evolution, are inspiring the design of novel materials with exceptional physical properties for high performance in adverse conditions. Their defining characteristics such as hierarchy, multifunctionality, and the capacity for self-healing, are currently being investigated.

The basic building blocks begin with the 20 amino acids and proceed to polypeptides, polysaccharides, and polypeptides–saccharides. These, in turn, compose the basic proteins, which are the primary constituents of the 'soft tissues' common to most biominerals. With well over 1000 proteins possible, current research emphasizes the use of collagen, chitin, keratin, and elastin. The 'hard' phases are often strengthened by crystalline minerals, which nucleate and grow in a biomediated environment that determines the size, shape and distribution of individual crystals. The most important mineral phases have been identified as hydroxyapatite, silica, and aragonite. Using the classification of Wegst and Ashby, the principal mechanical characteristics and structures of biological ceramics, polymer composites, elastomers, and cellular materials have been presented. Selected systems in each class are being investigated with emphasis on the relationship between their microstructure over a range of length scales and their mechanical response.

Thus, the crystallization of inorganic materials in nature generally occurs at ambient temperature and pressure. Yet the vital organisms through which these minerals form are capable of consistently producing extremely precise and complex structures. Understanding the processes in which living organisms control the growth of crystalline minerals such as silica could lead to significant advances in the field of materials science, and open the door to novel synthesis techniques for nanoscale composite materials, or nanocomposites.

High-resolution SEM observations were performed of the microstructure of the mother-of-pearl (or nacre) portion of the abalone shell. Those shells exhibit the highest mechanical strength and fracture toughness of any non-metallic substance known. The nacre from the shell of the abalone has become one of the more intensively studied biological structures in materials science. Clearly visible in these images are the neatly stacked (or ordered) mineral tiles separated by thin organic sheets along with a macrostructure of larger periodic growth bands which collectively form what scientists are currently referring to as a hierarchical composite structure. (The term hierarchy simply implies that there are a range of structural features which exist over a wide range of length scales).

The iridescent nacre inside a Nautilus shell.

Future developments reside in the synthesis of bio-inspired materials through processing methods and strategies that are characteristic of biological systems. These involve nanoscale self-assembly of the components and the development of hierarchical structures.

Polyethylene Terephthalate

Polyethylene terephthalate (sometimes written poly(ethylene terephthalate)), commonly abbreviated PET, PETE, or the obsolete PETP or PET-P, is the most common thermoplastic polymer resin of the polyester family and is used in fibers for clothing, containers for liquids and foods, thermoforming for manufacturing, and in combination with glass fiber for engineering resins.

It may also be referred to by the brand name Dacron; in Britain, Terylene; or, in Russia and the former Soviet Union, Lavsan.

The majority of the world's PET production is for synthetic fibers (in excess of 60%), with bottle production accounting for about 30% of global demand.In the context of textile applications, PET is referred to by its common name, *polyester*, whereas the acronym *PET* is generally used in relation to packaging. Polyester makes up about 18% of world polymer production and is the fourth-most-produced polymer; polyethylene (PE), polypropylene (PP) and polyvinyl chloride (PVC) are first, second and third, respectively.

PET consists of polymerized units of the monomer ethylene terephthalate, with repeating $(C_{10}H_8O_4)$ units. PET is commonly recycled, and has the number *1* as its recycling symbol.

Depending on its processing and thermal history, polyethylene terephthalate may exist both as an amorphous (transparent) and as a semi-crystalline polymer. The semicrystalline material might appear transparent (particle size < 500 nm) or opaque and white (particle size up to a few micrometers) depending on its crystal structure and particle size. Its monomer bis(2-hydroxyethyl) terephthalate can be synthesized by the esterification reaction between terephthalic acid and ethylene glycol with water as a byproduct, or by transesterification reaction between ethylene glycol and dimethyl terephthalate with methanol as a byproduct. Polymerization is through a polycondensation reaction of the monomers (done immediately after esterification/transesterification) with water as the byproduct.

Young's modulus (E)	2800–3100 MPa
Tensile strength(σ_t)	55–75 MPa
Elastic limit	50–150%
notch test	3.6 kJ/m²
Glass transition temperature (Tg)	67 to 81 °C
Vicat B	82 °C
linear expansion coefficient (α)	7×10^{-5}/K
Water absorption (ASTM)	0.16
Source	

Uses

Because PET is an excellent water and moisture barrier material, plastic bottles made from PET are widely used for soft drinks. For certain specialty bottles, such as those des-ignated for beer containment, PET sandwiches an additional polyvinyl alcohol (PVOH) layer to further reduce its oxygen permeability.

Biaxially oriented PET film (often known by one of its trade names, "Mylar") can be aluminized by evaporating a thin film of metal onto it to reduce its permeability, and to make it reflective and opaque (MPET). These properties are useful in many applications, including flexible food packaging and thermal insulation. See: "space blankets". Because of its high mechanical strength, PET film is often used in tape applications, such as the carrier for magnetic tape or backing for pressure-sensitive adhesive tapes.

Non-oriented PET sheet can be thermoformed to make packaging trays and blister packs. If crystallizable PET is used, the trays can be used for frozen dinners, since they withstand both freezing and oven baking temperatures. As opposed to amorphous PET, which is transparent, crystallizable PET or CPET tends to be black in colour.

When filled with glass particles or fibres, it becomes significantly stiffer and more durable.

PET is also used as a substrate in thin film solar cells.

Terylene (a trademark formed by inversion of (polyeth)ylene ter(ephthalate)) is also spliced into bell rope tops to help prevent wear on the ropes as they pass through the ceiling.

History

PET was patented in 1941 by John Rex Whinfield, James Tennant Dickson and their employer the Calico Printers' Association of Manchester, England. E. I. DuPont de Nemours in Delaware, USA, first used the trademark Mylar in June 1951 and received registration of it in 1952. It is still the best-known name used for polyester film. The current owner of the trademark is DuPont Teijin Films US, a partnership with a Japanese company.

In the Soviet Union, PET was first manufactured in the laboratories of the Institute of High-Molecular Compounds of the USSR Academy of Sciences in 1949, and its name "Lavsan" is an acronym thereof (**ла**боратории Института **в**ысокомолекулярных **с**оединений **А**кадемии **н**аук СССР).

The PET bottle was patented in 1973 by Nathaniel Wyeth.

Physical Properties

PET in its natural state is a colorless, semi-crystalline resin. Based on how it is processed, PET can be semi-rigid to rigid, and it is very lightweight. It makes a good gas and fair moisture barrier, as well as a good barrier to alcohol (requires additional "barrier" treatment) and solvents. It is strong and impact-resistant. PET becomes white when exposed to chloroform and also certain other chemicals such as toluene.

About 60% crystallization is the upper limit for commercial products, with the exception of polyester fibers. Clear products can be produced by rapidly cooling molten polymer below T_g glass transition temperature to form an amorphous solid. Like glass, amorphous PET forms when its molecules are not given enough time to arrange themselves in an orderly, crystalline fashion as the melt is cooled. At room temperature the molecules are frozen in place, but, if enough heat energy is put back into them by heating above T_g, they begin to move again, allowing crystals to nucleate and grow. This procedure is known as solid-state crystallization.

Sailcloth is typically made from PET fibers also known as polyester or under the brand name Dacron; colorful lightweight spinnakers are usually made of nylon.

When allowed to cool slowly, the molten polymer forms a more crystalline material. This material has spherulites containing many small crystallites when crystallized from an amorphous solid, rather than forming one large single crystal. Light tends to scatter as it crosses the boundaries between crystallites and the amorphous regions between them. This scattering means that crystalline PET is opaque and white in most cases. Fiber drawing is among the few industrial processes that produce a nearly single-crystal product.

Intrinsic Viscosity

One of the most important characteristics of PET is referred to as intrinsic viscosity (IV).

The intrinsic viscosity of the material, found by extrapolating to zero concentration of relative viscosity to concentration which is measured in deciliters per gram (dℓ/g). Intrinsic viscosity is dependent upon the length of its polymer chains but has no units due to being extrapolated to zero concentration. The longer the polymer chains the more entanglements between chains and therefore the higher the viscosity. The average chain length of a particular batch of resin can be controlled during polycondensation.

The intrinsic viscosity range of PET:

Fiber grade

     0.40–0.70 Textile

     0.72–0.98 Technical, tire cord

Film grade

     0.60–0.70 BoPET (biaxially oriented PET film)

     0.70–1.00 Sheet grade for thermoforming

Bottle grade

     0.70–0.78 Water bottles (flat)

     0.78–0.85 Carbonated soft drink grade

Monofilament, engineering plastic

     1.00–2.00

Drying

PET is hygroscopic, meaning that it absorbs water from its surroundings. However, when this "damp" PET is then heated, the water hydrolyzes the PET, decreasing its resilience. Thus, before the resin can be processed in a molding machine, it must be dried. Drying is achieved through the use of a desiccant or dryers before the PET is fed into the processing equipment.

Inside the dryer, hot dry air is pumped into the bottom of the hopper containing the resin so that it flows up through the pellets, removing moisture on its way. The hot wet air leaves the top of the hopper and is first run through an after-cooler, because it is easier to remove moisture from cold air than hot air. The resulting cool wet air is then passed through a desiccant bed. Finally, the cool dry air leaving the desiccant bed is re-heated in a process heater and sent back through the same processes in a closed loop. Typically, residual moisture levels in the resin must be less than 50 parts per million (parts of water per million parts of resin, by weight) before processing. Dryer residence time should not be shorter than about four hours. This is because drying the material in less than 4 hours would require a temperature above 160 °C, at which level hydrolysis would begin inside the pellets before they could be dried out.

PET can also be dried in compressed air resin dryers. Compressed air dryers do not reuse drying air. Dry, heated compressed air is circulated through the PET pellets as in the desiccant dryer, then released to the atmosphere.

Copolymers

In addition to pure (homopolymer) PET, PET modified by copolymerization is also available.

In some cases, the modified properties of copolymer are more desirable for a particular application. For example, cyclohexane dimethanol (CHDM) can be added to the polymer backbone in place of ethylene glycol. Since this building block is much larger (6 additional carbon atoms) than the ethylene glycol unit it replaces, it does not fit in with the neighboring chains the way an ethylene glycol unit would. This interferes with crystallization and lowers the polymer's melting temperature. In general, such PET is known as PETG or PET-G (Polyethylene terephthalate glycol-modified; Eastman Chemical, SK Chemicals, and Artenius Italia are some PETG manufacturers). PETG is a clear amorphous thermoplastic that can be injection molded or sheet extruded. It can be colored during processing.

Another common modifier is isophthalic acid, replacing some of the 1,4-(*para-*) linked terephthalate units. The 1,2-(*ortho-*) or 1,3-(*meta-*) linkage produces an angle in the chain, which also disturbs crystallinity.

Such copolymers are advantageous for certain molding applications, such as thermoforming, which is used for example to make tray or blister packaging from co-PET film, or amorphous PET sheet (A-PET) or PETG sheet. On the other hand, crystallization is important in other applications where mechanical and dimensional stability are important, such as seat belts. For PET bottles, the use of small amounts of isophthalic acid, CHDM, diethylene glycol (DEG) or other comonomers can be useful: if only small amounts of comonomers are used, crystallization is slowed but not prevented entirely. As a result, bottles are obtainable via stretch blow molding ("SBM"), which are both clear and crystalline enough to be an adequate barrier to aromas and even gases, such as carbon dioxide in carbonated beverages.

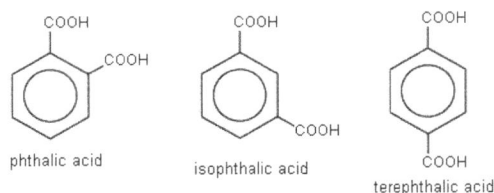

Replacing terephthalic acid (right) with isophthalic acid (center) creates a kink in the PET chain, interfering with crystallization and lowering the polymer's melting point.

Production

Polyethylene terephthalate is produced from ethylene glycol and dimethyl terephthalate ($C_6H_4(CO_2CH_3)_2$) or terephthalic acid.

The former is a transesterification reaction, whereas the latter is an esterification reaction.

Dimethyl Terephthalate Process

In dimethyl terephthalate process, this compound and excess ethylene glycol are reacted in the melt at 150–200 °C with a basic catalyst. Methanol (CH_3OH) is removed by distillation to drive the reaction forward. Excess ethylene glycol is distilled off at higher temperature with the aid of vacuum. The second transesterification step proceeds at 270–280 °C, with continuous distillation of ethylene glycol as well.

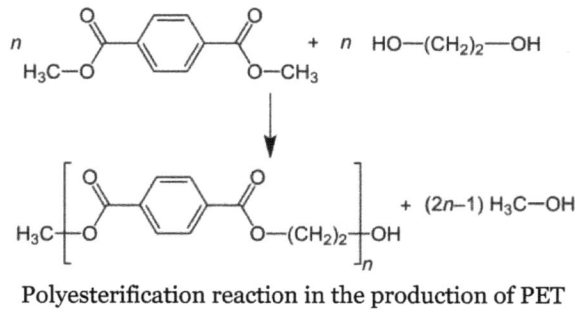

Polyesterification reaction in the production of PET

The reactions are idealized as follows:

First step

$$C_6H_4(CO_2CH_3)_2 + 2\ HOCH_2CH_2OH\quad C_6H_4(CO_2CH_2CH_2OH)_2 + 2\ CH_3OH$$

Second step

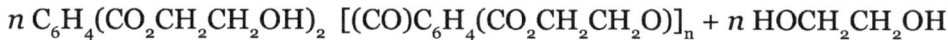

$$n\ C_6H_4(CO_2CH_2CH_2OH)_2\quad [(CO)C_6H_4(CO_2CH_2CH_2O)]_n + n\ HOCH_2CH_2OH$$

Terephthalic Acid Process

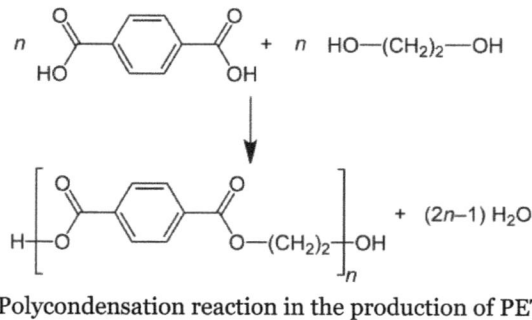

Polycondensation reaction in the production of PET

In the terephthalic acid process, esterification of ethylene glycol and terephthalic acid is conducted directly at moderate pressure (2.7–5.5 bar) and high temperature (220–260 °C). Water is eliminated in the reaction, and it is also continuously removed by distillation:

$$n\ C_6H_4(CO_2H)_2 + n\ HOCH_2CH_2OH\quad [(CO)C_6H_4(CO_2CH_2CH_2O)]_n + 2n\ H_2O$$

Degradation

PET is subjected to various types of degradations during processing. The main degradations that can occur are hydrolytic, and probably most important, thermal oxidation. When PET degrades, several things happen: discoloration, chain scissions resulting in reduced molecular weight, formation of acetaldehyde, and cross-links ("gel" or "fish-eye" formation). Discoloration is due to the formation of various chromophoric systems following prolonged thermal treatment at elevated temperatures. This becomes a problem when the optical requirements of the polymer are very high, such as in packaging applications. The thermal and thermooxidative degradation results in poor processibility characteristics and performance of the material.

One way to alleviate this is to use a copolymer. Comonomers such as CHDM or isophthalic acid lower the melting temperature and reduce the degree of crystallinity of PET (especially important when the material is used for bottle manufacturing). Thus, the resin can be plastically formed at lower temperatures and/or with lower force. This helps to prevent degradation, reducing the acetaldehyde content of the finished product to an acceptable (that is, unnoticeable) level. See copolymers, above. Another way to improve the stability of the polymer is to use stabilizers, mainly antioxidants such as phosphites. Recently, molecular level stabilization of the material using nanostructured chemicals has also been considered.

Acetaldehyde

Acetaldehyde is a colorless, volatile substance with a fruity smell. Although it forms naturally in some fruit, it can cause an off-taste in bottled water. Acetaldehyde forms by degradation of PET through the mishandling of the material. High temperatures (PET decomposes above 300 °C or 570 °F), high pressures, extruder speeds (excessive shear flow raises temperature), and long barrel residence times all contribute to the production of acetaldehyde. When acetaldehyde is produced, some of it remains dissolved in the walls of a container and then diffuses into the product stored inside, altering the taste and aroma. This is not such a problem for non-consumables (such as shampoo), for fruit juices (which already contain acetaldehyde), or for strong-tasting drinks like soft drinks. For bottled water, however, low acetaldehyde content is quite important, because, if nothing masks the aroma, even extremely low concentrations (10–20 parts per billion in the water) of acetaldehyde can produce an off-taste.

Antimony

Antimony (Sb) is a metalloid element that is used as a catalyst in the form of compounds such as antimony trioxide (Sb_2O_3) or antimony triacetate in the production of PET. After manufacturing, a detectable amount of antimony can be found on the surface of the product. This residue can be removed with washing. Antimony also remains in the material itself and can, thus, migrate out into food and drinks. Exposing PET to boiling or microwaving can increase the levels of antimony significantly, possibly above USEPA maximum contamination levels. The drinking water limit assessed by WHO is 20 parts per billion (WHO, 2003), and the drinking water limit in the USA is 6 parts per billion. Although antimony trioxide is of low toxicity when taken orally, its presence is still of concern. The Swiss Federal Office of Public Health investigated the amount of antimony migration, comparing waters bottled in PET and glass: The antimony concentrations of the water in PET bottles were higher, but still well below the allowed maximum concentration. The Swiss Federal Office of Public Health concluded that small amounts of antimony migrate from the PET into bottled water, but that the health risk of the resulting low concentrations is negligible (1% of the "tolerable daily intake" determined by the WHO). A later (2006) but more widely publicized study found similar amounts of antimony in water in PET bottles. The WHO has published a risk assessment for antimony in drinking water.

Fruit juice concentrates (for which no guidelines are established), however, that were produced and bottled in PET in the UK were found to contain up to 44.7 µg/L of antimony, well above the EU limits for tap water of 5 µg/L.

Biodegradation

Nocardia can degrade PET with an esterase enzyme.

Japanese scientists have isolated a bacterium *Ideonella sakaiensis* that possesses two enzymes which can break down the PET into smaller pieces that the bacterium can digest. A colony of *I. sakaiensis* can disintegrate a plastic film in about six weeks.

Safety

Commentary published in *Environmental Health Perspectives* in April 2010 suggested that PET might yield endocrine disruptors under conditions of common use and recommended research on this topic. Proposed mechanisms include leaching of phthalates as well as leaching of antimony. Article published in *Journal of Environmental Monitoring* in April 2012 concludes that antimony concentration in deionized water stored in PET bottles stays within EU's acceptable limit even if stored briefly at temperatures up to 60 °C (140 °F), while bottled contents (water or soft drinks) may occasionally exceed the EU limit after less than a year of storage at room temperature.

Bottle Processing Equipment

There are two basic molding methods for PET bottles, one-step and two-step. In two-step molding, two separate machines are used. The first machine injection molds the preform, which resembles a test tube, with the bottle-cap threads already molded into place. The body of the tube is significantly thicker, as it will be inflated into its final shape in the second step using stretch blow molding.

A finished PET drink bottle compared to the preform from which it is made.

In the second step, the preforms are heated rapidly and then inflated against a two-part mold to form them into the final shape of the bottle. Preforms (uninflated bottles) are now also used as robust and unique containers themselves; besides novelty candy, some Red Cross chapters distribute them as part of the Vial of Life program to homeowners to store medical history for emergency responders. Another increasingly common use for the preforms are containers in the outdoor activity geocaching.

In one-step machines, the entire process from raw material to finished container is conducted within one machine, making it especially suitable for molding non-standard shapes (custom molding), including jars, flat oval, flask shapes, etc. Its greatest merit is the reduction in space, product handling and energy, and far higher visual quality than can be achieved by the two-step system.

Polyester Recycling Industry

While most thermoplastics can, in principle, be recycled, PET bottle recycling is more practical than many other plastic applications because of the high value of the resin and the almost exclusive use of PET for widely used water and carbonated soft drink bottling. PET has a resin identification code of 1. The prime uses for recycled PET are polyester fiber, strapping, and non-food containers.

PETE

In year 2016, it was estimated that 56 million tons of PET are produced each year.

Because of the recyclability of PET and the relative abundance of post-consumer waste in the form of bottles, PET is rapidly gaining market share as a carpet fiber. Mohawk Industries released ever-STRAND in 1999, a 100% post-consumer recycled content PET fiber. Since that time, more than 17 billion bottles have been recycled into carpet fiber. Pharr Yarns, a supplier to numerous carpet manufacturers including Looptex, Dobbs Mills, and Berkshire Flooring, produces a BCF (bulk continuous filament) PET carpet fiber containing a minimum of 25% post-consumer recycled content.

PET, as with many plastics, is also an excellent candidate for thermal disposal (incineration), as it is composed of carbon, hydrogen, and oxygen, with only trace amounts of catalyst elements (but no sulfur). PET has the energy content of soft coal.

When recycling polyethylene terephthalate or PET or polyester, in general two ways have to be differentiated:

1. The chemical recycling back to the initial raw materials purified terephthalic acid (PTA) or dimethyl terephthalate (DMT) and ethylene glycol (EG) where the polymer structure is destroyed completely, or in process intermediates like bis(2-hydroxyethyl) terephthalate

2. The mechanical recycling where the original polymer properties are being maintained or reconstituted.

Chemical recycling of PET will become cost-efficient only applying high capacity recycling lines of more than 50,000 tons/year. Such lines could only be seen, if at all, within the production sites of very large polyester producers. Several attempts of industrial magnitude to establish such chemical recycling plants have been made in the past but without resounding success. Even the promising chemical recycling in Japan has not become an industrial breakthrough so far. The two reasons for this are: at first, the difficulty of consistent and continuous waste bottles sourcing in such a huge amount at one single site, and, at second, the steadily increased prices and price volatility of collected bottles. The prices of baled bottles increased for instance between the years 2000 and 2008 from about 50 Euro/ton to over 500 Euro/ton in 2008.

Mechanical recycling or direct circulation of PET in the polymeric state is operated in most diverse variants today. These kinds of processes are typical of small and medium-size industry. Cost-efficiency can already be achieved with plant capacities within a range of 5000–20,000 tons/year. In this case, nearly all kinds of recycled-material feedback into the material circulation are possible today. These diverse recycling processes are being discussed hereafter in detail.

Besides chemical contaminants and degradation products generated during first processing and usage, mechanical impurities are representing the main part of quality depreciating impurities in the recycling stream. Recycled materials are increasingly introduced into manufacturing processes, which were originally designed for new materials only. Therefore, efficient sorting, separation and cleaning processes become most important for high quality recycled polyester.

When talking about polyester recycling industry, we are concentrating mainly on recycling of PET bottles, which are meanwhile used for all kinds of liquid packaging like water, carbonated soft drinks, juices, beer, sauces, detergents, household chemicals and so on. Bottles are easy to distinguish because of shape and consistency and separate from waste plastic streams either by automatic or by hand-sorting processes. The established polyester recycling industry consists of three major sections:

- PET bottle collection and waste separation: waste logistics

- Production of clean bottle flakes: flake production

- Conversion of PET flakes to final products: flake processing

Intermediate product from the first section is baled bottle waste with a PET content greater than 90%. Most common trading form is the bale but also bricked or even loose, pre-cut bottles are common in the market. In the second section, the collected bottles are converted to clean PET bottle flakes. This step can be more or less complex and complicated depending on required final flake quality. During the third step, PET bottle flakes are processed to any kind of products like film, bottles, fiber, filament, strapping or intermediates like pellets for further processing and engineering plastics.

Besides this external (post-consumer) polyester bottle recycling, numbers of internal (pre-consumer) recycling processes exist, where the wasted polymer material does not exit the production site to the free market, and instead is reused in the same production circuit. In this way, fiber waste is directly reused to produce fiber, preform waste is directly reused to produce preforms, and film waste is directly reused to produce film.

PET Bottle Recycling

Purification and Decontamination

The success of any recycling concept is hidden in the efficiency of purification and decontamination at the right place during processing and to the necessary or desired extent.

In general, the following applies: The earlier in the process foreign substances are removed, and the more thoroughly this is done, the more efficient the process is.

Workers sort an incoming stream of various plastics, mixed with some pieces of un-recyclable litter.

The high plasticization temperature of PET in the range of 280 °C (536 °F) is the reason why almost all common organic impurities such as PVC, PLA, polyolefin, chemical wood-pulp and paper fibers, polyvinyl acetate, melt adhesive, coloring agents, sugar, and protein residues are transformed into colored degradation products that, in their turn, might release in addition reactive degradation products.Then, the number of defects in the polymer chain increases considerably. The particle size distribution of impurities is very wide, the big particles of 60–1000 µm—which are visible by naked eye and easy to filter—representing the lesser evil, since their total surface is relatively small and the degradation speed is therefore lower. The influence of the microscopic particles, which —because they are many— increase the frequency of defects in the polymer, is relatively greater.

Bales of crushed blue PET bottles.

The motto "What the eye does not see the heart cannot grieve over" is considered to be very important in many recycling processes. Therefore, besides efficient sorting, the removal of visible impurity particles by melt filtration processes plays a particular part in this case.

Bales of crushed PET bottles sorted according to color: green, transparent, and blue.

In general, one can say that the processes to make PET bottle flakes from collected bottles are as versatile as the different waste streams are different in their composition and quality. In view of technology there isn't just one way to do it. Meanwhile, there are many engineering companies that are offering flake production plants and components, and it is difficult to decide for one or other plant design. Nevertheless, there are processes that are sharing most of these principles. Depending on composition and impurity level of input material, the general following process steps are applied.

- Bale opening, briquette opening
- Sorting and selection for different colors, foreign polymers especially PVC, foreign matter, removal of film, paper, glass, sand, soil, stones, and metals
- Pre-washing without cutting
- Coarse cutting dry or combined to pre-washing
- Removal of stones, glass, and metal
- Air sifting to remove film, paper, and labels
- Grinding, dry and / or wet
- Removal of low-density polymers (cups) by density differences
- Hot-wash
- Caustic wash, and surface etching, maintaining intrinsic viscosity and decontamination
- Rinsing
- Clean water rinsing
- Drying
- Air-sifting of flakes
- Automatic flake sorting
- Water circuit and water treatment technology
- Flake quality control

Impurities and Material Defects

The number of possible impurities and material defects that accumulate in the polymeric material is increasing permanently—when processing as well as when using polymers—taking into account a growing service lifetime, growing final applications and repeated recycling. As far as recycled PET bottles are concerned, the defects mentioned can be sorted in the following groups:

- Reactive polyester OH- or COOH- end groups are transformed into dead or non-reactive end groups, e.g. formation of vinyl ester end groups through dehydration or decarboxylation of terephthalate acid, reaction of the OH- or COOH- end groups with mono-functional degradation products like mono-carbonic acids or alcohols. Results are decreased reactivity during re-polycondensation or re-SSP and broadening the molecular weight distribution.

- The end group proportion shifts toward the direction of the COOH end groups built up through a thermal and oxidative degradation. The results are decrease in reactivity, and increase in the acid autocatalytic decomposition during thermal treatment in presence of humidity.

- Number of polyfunctional macromolecules increases. Accumulation of gels and long-chain branching defects.

- Number, concentration, and variety of nonpolymer-identical organic and inorganic foreign substances are increasing. With every new thermal stress, the organic foreign substances will react by decomposition. This is causing the liberation of further degradation-supporting substances and coloring substances.

- Hydroxide and peroxide groups build up at the surface of the products made of polyester in presence of air (oxygen) and humidity. This process is accelerated by ultraviolet light. During an ulterior treatment process, hydro peroxides are a source of oxygen radicals, which are source of oxidative degradation. Destruction of hydro peroxides is to happen before the first thermal treatment or during plasticization and can be supported by suitable additives like antioxidants.

Taking into consideration the above-mentioned chemical defects and impurities, there is an ongoing modification of the following polymer characteristics during each recycling cycle, which are detectable by chemical and physical laboratory analysis.
In Particular:

- Increase of COOH end-groups

- Increase of color number b

- Increase of haze (transparent products)

- Increase of oligomer content

- Reduction in filterability

- Increase of by-products content such as acetaldehyde, formaldehyde

- Increase of extractable foreign contaminants

- Decrease in color L

- Decrease of intrinsic viscosity or dynamic viscosity

- Decrease of crystallization temperature and increase of crystallization speed

- Decrease of the mechanical properties like tensile strength, elongation at break or elastic modulus

- Broadening of molecular weight distribution

The recycling of PET-bottles is meanwhile an industrial standard process that is offered by a wide variety of engineering companies.

Processing Examples for Recycled Polyester

Recycling processes with polyester are almost as varied as the manufacturing processes based on primary pellets or melt. Depending on purity of the recycled materials, polyester can be used today in most of the polyester manufacturing processes as blend with virgin polymer or increasingly as 100% recycled polymer. Some exceptions like BOPET-film of low thickness, special applications like optical film or yarns through FDY-spinning at > 6000 m/min, microfilaments, and micro-fibers are produced from virgin polyester only.

Simple re-pelletizing of Bottle Flakes

This process consists of transforming bottle waste into flakes, by drying and crystallizing the flakes, by plasticizing and filtering, as well as by pelletizing. Product is an amorphous re-granulate of an intrinsic viscosity in the range of 0.55–0.7 dℓ/g, depending on how complete pre-drying of PET flakes has been done.

Special feature are: Acetaldehyde and oligomers are contained in the pellets at lower level; the viscosity is reduced somehow, the pellets are amorphous and have to be crystallized and dried before further processing.

Processing to:

- A-PET film for thermoforming

- Addition to PET virgin production

- BoPET packaging film

- PET Bottle resin by SSP

- Carpet yarn

- Engineering plastic

- Filaments

- Non-woven

- Packaging stripes

- Staple fibre.

Choosing the re-pelletizing way means having an additional conversion process that is, at the one side, energy-intensive and cost-consuming, and causes thermal destruction. At the other side, the pelletizing step is providing the following advantages:

- Intensive melt filtration

- Intermediate quality control

- Modification by additives

- Product selection and separation by quality

- Processing flexibility increased

- Quality uniformization.

Manufacture of PET-pellets or Flakes for Bottles (Bottle to Bottle) and A-PET

This process is, in principle, similar to the one described above; however, the pellets produced are directly (continuously or discontinuously) crystallized and then subjected to a solid-state poly-condensation (SSP) in a tumbling drier or a vertical tube reactor. During this processing step, the corresponding intrinsic viscosity of 0.80–0.085 dℓ/g is rebuild again and, at the same time, the acetaldehyde content is reduced to < 1 ppm.

The fact that some machine manufacturers and line builders in Europe and USA make efforts to offer independent recycling processes, e.g. the so-called bottle-to-bottle (B-2-B) process, such as BePET, Starlinger, URRC or BÜHLER, aims at generally furnishing proof of the "existence" of the required extraction residues and of the removal of model contaminants according to FDA applying the so-called challenge test, which is necessary for the application of the treated polyester in the food sector. Besides this process approval it is nevertheless necessary that any user of such processes has to constantly check the FDA-limits for the raw materials manufactured by himself for his process.

Direct Conversion of Bottle Flakes

In order to save costs, an increasing number of polyester intermediate producers like spinning mills, strapping mills, or cast film mills are working on the direct use of the PET-flakes, from the treatment of used bottles, with a view to manufacturing an increasing number of polyester inter-mediates. For the adjustment of the necessary viscosity, besides an efficient drying of the flakes, it is possibly necessary to also reconstitute the viscosity through polycondensation in the melt phase or solid-state polycondensation of the flakes. The latest PET flake conversion processes are applying twin screw extruders, multi-screw extruders or multi-rotation systems and coincidental vacuum degassing to remove moisture and avoid flake pre-drying. These processes allow the con-version of undried PET flakes without substantial viscosity decrease caused by hydrolysis.

With regard to the consumption of PET bottle flakes, the main portion of about 70% is converted to fibers and filaments. When using directly secondary materials such as bottle flakes in spinning processes, there are a few processing principles to obtain.

High-speed spinning processes for the manufacture of POY normally need a viscosity of 0.62–0.64 dℓ/g. Starting from bottle flakes, the viscosity can be set via the degree of drying. The additional use of TiO_2 is necessary for full dull or semi dull yarn. In order to protect the spinnerets, an efficient filtration of the melt is, in any case is necessary. For the time-being, the amount of POY made of 100% recycling polyester is rather low because this process requires high purity of spinning melt. Most of the time, a blend of virgin and recycled pellets is used.

Staple fibers are spun in an intrinsic viscosity range that lies rather somewhat lower and that should be between 0.58 and 0.62 dℓ/g. In this case, too, the required viscosity can be adjusted via drying or vacuum adjustment in case of vacuum extrusion. For adjusting the viscosity, however, an addition of chain length modifier like ethylene glycol or diethylene glycol can also be used.

Spinning non-woven—in the fine titer field for textile applications as well as heavy spinning non-woven as basic materials, e.g. for roof covers or in road building—can be manufactured by spinning bottle flakes. The spinning viscosity is again within a range of 0.58–0.65 dℓ/g.

One field of increasing interest where recycled materials are used is the manufacture of high-tenacity packaging stripes, and monofilaments. In both cases, the initial raw material is a mainly recycled material of higher intrinsic viscosity. High-tenacity packaging stripes as well as monofilament are then manufactured in the melt spinning process.

Recycling to the Monomers

Polyethylene terephthalate can be depolymerized to yield the constituent monomers. After purification, the monomers can be used to prepare new polyethylene terephthalate. The ester bonds in polyethylene terephthalate may be cleaved by hydrolysis, or by transesterification. The reactions are simply the reverse of those used in production.

Partial Glycolysis

Partial glycolysis (transesterification with ethylene glycol) converts the rigid polymer into short-chained oligomers that can be melt-filtered at low temperature. Once freed of the impurities, the oligomers can be fed back into the production process for polymerization.

The task consists in feeding 10–25% bottle flakes while maintaining the quality of the bottle pellets that are manufactured on the line. This aim is solved by degrading the PET bottle flakes—already during their first plasticization, which can be carried out in a single- or multi-screw extruder—to an intrinsic viscosity of about 0.30 dℓ/g by adding small quantities of ethylene glycol and by subjecting the low-viscosity melt stream to an efficient filtration directly after plasticization. Furthermore, temperature is brought to the lowest possible limit. In addition, with this way of processing, the possibility of a chemical decomposition of the hydro peroxides is possible by adding a corresponding P-stabilizer directly when plasticizing. The destruction of the hydro peroxide groups is, with other processes, already carried out during the last step of flake treatment for instance by adding H_3PO_3. The partially glycolyzed and finely filtered recycled material is continuously fed to the esterification or prepolycondensation reactor, the dosing quantities of the raw materials are being adjusted accordingly.

Total Glycolysis, Methanolysis, and Hydrolysis

The treatment of polyester waste through total glycolysis to fully convert the polyester to bis(2-hydroxyethyl) terephthalate ($C_6H_4(CO_2CH_2CH_2OH)_2$). This compound is purified by vacuum distillation, and is one of the intermediates used in polyester manufacture . The reaction involved is as follows:

$$[(CO)C_6H_4(CO_2CH_2CH_2O)]_n + n \text{ } HOCH_2CH_2OH \text{ } n \text{ } C_6H_4(CO_2CH_2CH_2OH)_2$$

This recycling route has been executed on an industrial scale in Japan as experimental production.

Similar to total glycolysis, methanolysis converts the polyester to dimethyl terephthalate, which can be filtered and vacuum distilled:

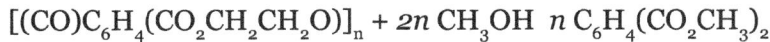

$$[(CO)C_6H_4(CO_2CH_2CH_2O)]_n + 2n \text{ } CH_3OH \text{ } n \text{ } C_6H_4(CO_2CH_3)_2$$

Methanolysis is only rarely carried out in industry today because polyester production based on dimethyl terephthalate has shrunk tremendously, and many dimethyl terephthalate producers have disappeared.

Also as above, polyethylene terephthalate can be hydrolyzed to terephthalic acid and ethylene glycol under high temperature and pressure. The resultant crude terephthalic acid can be purified by recrystallization to yield material suitable for re-polymerization:

$$[(CO)C_6H_4(CO_2CH_2CH_2O)]_n + 2n \text{ } H_2O \text{ } n \text{ } C_6H_4(CO_2H)_2 + n \text{ } HOCH_2CH_2OH$$

This method does not appear to have been commercialized yet.

References

- Harris, D.C., "Materials for Infrared Windows and Domes: Properties and Performance", SPIE PRESS Monograph, Vol. PM70 (Int. Society of Optical Engineers, Bellingham WA, 2009) ISBN 978-0-8194-5978-7

- Brinker, C.J.; Scherer, G.W. (1990). Sol-Gel Science: The Physics and Chemistry of Sol-Gel Processing. Academic Press. ISBN 0-12-134970-5.

- J. G. Speight, Norbert Adolph Lange (2005). McGraw-Hill, ed. Lange's handbook of chemistry (16 ed.). pp. 2.807–2.758. ISBN 0-07-143220-5.

- Ulrich K. Thiele (2007) Polyester Bottle Resins, Production, Processing, Properties and Recycling, Heidelberg, Germany, pp. 85 ff, ISBN 978-3-9807497-4-9

- V.B. Gupta and Z. Bashir (2002) Chapter 7, p. 320 in Stoyko Fakirov (ed.) Handbook of Thermoplastic Polyesters, Wiley-VCH, Weinheim, ISBN 3-527-30113-5.

- PET-Recycling Forum; "Current Technological Trends in Polyester Recycling"; 9th International Polyester Recycling Forum Washington, 2006; São Paulo; ISBN 3-00-019765-6

- Ulrich K. Thiele (2007) Polyester Bottle Resins Production, Processing, Properties and Recycling, PETplanet Publisher GmbH, Heidelberg, Germany, pp. 259 ff, ISBN 978-3-9807497-4-9

- "Could a new plastic-eating bacteria help combat this pollution scourge?". The Guardian. 10 March 2016. Retrieved 11 March 2016.

⁵

Applications of Nanocomposites

Nanocomposites are used in a variety of devices and technologies that provide more efficiency and longer-life. Some of these applications that have been listed in this chapter are light-emitting diode, photodiode, solar film, colloid, gel etc. Nanocomposites are best understood in confluence with the major topics listed in the following chapter.

Light-emitting Diode

A light-emitting diode (LED) is a two-lead semiconductor light source. It is a p–n junction diode, which emits light when activated. When a suitable voltage is applied to the leads, electrons are able to recombine with electron holes within the device, releasing energy in the form of photons. This effect is called electroluminescence, and the color of the light (corresponding to the energy of the photon) is determined by the energy band gap of the semiconductor.

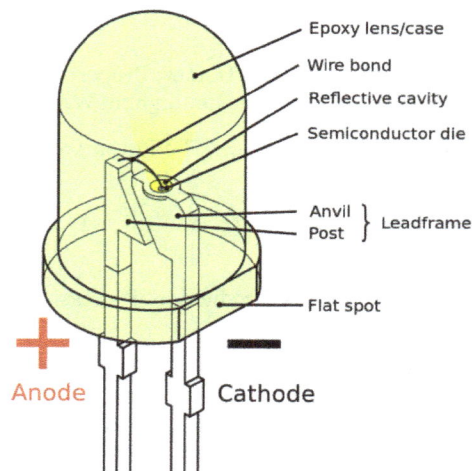

Parts of an LED. Although unlabeled, the flat bottom surfaces of the anvil and post embedded inside the epoxy act as anchors, to prevent the conductors from being forcefully pulled out via mechanical strain or vibration.

An LED is often small in area (less than 1 mm²) and integrated optical components may be used to shape its radiation pattern.

Appearing as practical electronic components in 1962, the earliest LEDs emitted low-intensity infrared light. Infrared LEDs are still frequently used as transmitting elements in remote-control circuits, such as those in remote controls for a wide variety of consumer electronics. The first visible-light LEDs were also of low intensity, and limited to red. Modern LEDs are available across the visible, ultraviolet, and infrared wavelengths, with very high brightness.

Early LEDs were often used as indicator lamps for electronic devices, replacing small incandescent bulbs. They were soon packaged into numeric readouts in the form of seven-segment displays, and were commonly seen in digital clocks.

A bulb-shaped modern retrofit LED lamp with aluminium heat sink, a light diffusing dome and E27 screw base, using a built-in power supply working on mains voltage

Recent developments in LEDs permit them to be used in environmental and task lighting. LEDs have many advantages over incandescent light sources including lower energy consumption, longer lifetime, improved physical robustness, smaller size, and faster switching. Light-emitting diodes are now used in applications as diverse as aviation lighting, automotive headlamps, advertising, general lighting, traffic signals, camera flashes and lighted wallpaper. As of 2015, LEDs powerful enough for room lighting remain somewhat more expensive, and require more precise current and heat management than compact fluorescent lamp sources of comparable output.

LEDs have allowed new text, video displays, and sensors to be developed, while their high switching rates are also used in advanced communications technology.

History

Discoveries and Early Devices

Electroluminescence as a phenomenon was discovered in 1907 by the British experimenter H. J. Round of Marconi Labs, using a crystal of silicon carbide and a cat's-whisker detector. Soviet inventor Oleg Losev reported creation of the first LED in 1927. His research was distributed in Soviet, German and British scientific journals, but no practical use was made of the discovery for several decades. Kurt Lehovec, Carl Accardo and Edward Jamgochian, explained these first light-emitting diodes in 1951 using an apparatus employing SiC crystals with a current source of battery or pulse generator and with a comparison to a variant, pure, crystal in 1953.

Green electroluminescence from a point contact on a crystal of SiC recreates Round's original experiment from 1907.

Rubin Braunstein of the Radio Corporation of America reported on infrared emission from gallium arsenide (GaAs) and other semiconductor alloys in 1955. Braunstein observed infrared emission generated by simple diode structures using gallium antimonide (GaSb), GaAs, indium phosphide (InP), and silicon-germanium (SiGe) alloys at room temperature and at 77 Kelvin.

In 1957, Braunstein further demonstrated that the rudimentary devices could be used for non-radio communication across a short distance. As noted by Kroemer Braunstein".. had set up a simple optical communications link: Music emerging from a record player was used via suitable electronics to modulate the forward current of a GaAs diode. The emitted light was detected by a PbS diode some distance away. This signal was fed into an audio amplifier, and played back by a loudspeaker. Intercepting the beam stopped the music. We had a great deal of fun playing with this setup." This setup presaged the use of LEDs for optical communication applications.

A Texas Instruments SNX-100 GaAs LED contained in a TO-18 transistor metal case.

In September 1961, while working at Texas Instruments in Dallas, Texas, James R. Biard and Gary Pittman discovered near-infrared (900 nm) light emission from a tunnel diode they had constructed on a GaAs substrate. By October 1961, they had demonstrated efficient light emission and signal coupling between a GaAs p-n junction light emitter and an electrically-isolated semiconductor photodetector. On August 8, 1962, Biard and Pittman filed a patent titled "Semiconductor Radiant Diode" based on their findings, which described a zinc diffused p–n junction LED with a spaced cathode contact to allow for efficient emission of infrared light under forward bias. After establishing the priority of their work based on engineering notebooks predating submissions from G.E. Labs, RCA Research Labs, IBM Research Labs, Bell Labs, and Lincoln Lab at MIT, the U.S. patent office issued the two inventors the patent for the GaAs infrared (IR) light-emitting diode (U.S. Patent US3293513), the first practical LED. Immediately after filing the patent, Texas Instruments (TI) began a project to manufacture infrared diodes. In October 1962, TI announced the first LED commercial product (the SNX-100), which employed a pure GaAs crystal to emit a 890 nm light output. In October 1963, TI announced the first commercial hemispherical LED, the SNX-110.

The first visible-spectrum (red) LED was developed in 1962 by Nick Holonyak, Jr., while working at General Electric. Holonyak first reported his LED in the journal *Applied Physics Letters* on the December 1, 1962. M. George Craford, a former graduate student of Holonyak, invented the first yellow LED and improved the brightness of red and red-orange LEDs by a factor of ten in 1972. In 1976, T. P. Pearsall created the first high-brightness, high-efficiency LEDs for optical fiber telecommunications by inventing new semiconductor materials specifically adapted to optical fiber transmission wavelengths.

Initial Commercial Development

The first commercial LEDs were commonly used as replacements for incandescent and neon indicator lamps, and in seven-segment displays, first in expensive equipment such as laboratory and electronics test equipment, then later in such appliances as TVs, radios, telephones, calculators, as well as watches. Until 1968, visible and infrared LEDs were extremely costly, in the order of US$200 per unit, and so had little practical use. The Monsanto Company was the first organization to mass-produce visible LEDs, using gallium arsenide phosphide (GaAsP) in 1968 to produce red LEDs suitable for indicators. Hewlett Packard (HP) introduced LEDs in 1968, initially using GaAsP supplied by Monsanto. These red LEDs were bright enough only for use as indicators, as the light output was not enough to illuminate an area. Readouts in calculators were so small that plastic lenses were built over each digit to make them legible. Later, other colors became widely available and appeared in appliances and equipment. In the 1970s commercially successful LED devices at less than five cents each were produced by Fairchild Optoelectronics. These devices employed compound semiconductor chips fabricated with the planar process invented by Dr. Jean Hoerni at Fairchild Semiconductor. The combination of planar processing for chip fabrication and innovative packaging methods enabled the team at Fairchild led by optoelectronics pioneer Thomas Brandt to achieve the needed cost reductions. These methods continue to be used by LED producers.

LED display of a TI-30 scientific calculator (ca. 1978), which uses plastic lenses to increase the visible digit size

Most LEDs were made in the very common 5 mm T1¾ and 3 mm T1 packages, but with rising power output, it has grown increasingly necessary to shed excess heat to maintain reliability, so more complex packages have been adapted for efficient heat dissipation. Packages for state-of-the-art high-power LEDs bear little resemblance to early LEDs.

Blue LED

Blue LEDs were first developed by Herbert Paul Maruska at RCA in 1972 using gallium nitride (GaN) on a sapphire substrate. SiC-types were first commercially sold in the United States by Cree in 1989. However, neither of these initial blue LEDs were very bright.

The first high-brightness blue LED was demonstrated by Shuji Nakamura of Nichia Corporation in 1994 and was based on InGaN. In parallel, Isamu Akasaki and Hiroshi Amano in Nagoya were working on developing the important GaN nucleation on sapphire substrates and the demonstration of p-type doping of GaN. Nakamura, Akasaki and Amano were awarded the 2014 Nobel prize in physics for their work. In 1995, Alberto Barbieri at the Cardiff University Laboratory (GB) investigated the efficiency and reliability of high-brightness LEDs and demonstrated a "transparent contact" LED using indium tin oxide (ITO) on (AlGaInP/GaAs).

In 2001 and 2002, processes for growing gallium nitride (GaN) LEDs on silicon were successfully demonstrated. In January 2012, Osram demonstrated high-power InGaN LEDs grown on silicon substrates commercially.

White LEDs and The Illumination Breakthrough

The attainment of high efficiency in blue LEDs was quickly followed by the development of the first white LED. In this device a Y_3Al_5O 12:Ce (known as "YAG") phosphor coating on the emitter absorbs some of the blue emission and produces yellow light through fluorescence. The combination of that yellow with remaining blue light appears white to the eye. However, using different phosphors (fluorescent materials) it also became possible to instead produce green and red light through fluorescence. The resulting mixture of red, green and blue is not only perceived by humans as white light but is superior for illumination in terms of color rendering, whereas one cannot appreciate the color of red or green objects illuminated only by the yellow (and remaining blue) wavelengths from the YAG phosphor.

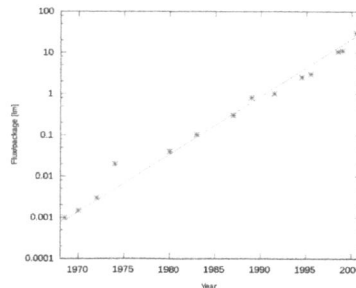

Illustration of Haitz's law, showing improvement in light output per LED over time, with a logarithmic scale on the vertical axis

The first white LEDs were expensive and inefficient. However, the light output of LEDs has increased exponentially, with a doubling occurring approximately every 36 months since the 1960s (similar to Moore's law). This trend is generally attributed to the parallel development of other semiconductor technologies and advances in opticsand materials science, and has been called Haitz's law after Dr. Roland Haitz.

The light output and efficiency of blue and near-ultraviolet LEDs rose as the cost of reliable devices fell: this led to the use of (relatively) high-power white-light LEDs for the purpose of illumination which are replacing incandescent and fluorescent lighting.

White LEDs can now produce over 300 lumens per watt of electricity while lasting up to 100,000 hours. Compared to incandescent bulbs, this is not only a huge increase in electrical efficiency, but - over time - a similar or lower cost per bulb.

Working Principle

A P-N junction can convert absorbed light energy into a proportional electric current. The same process is reversed here (i.e. the P-N junction emits light when electrical energy is applied to it). This phenomenon is generally called electroluminescence, which can be defined as the emission of light from a semi-conductor under the influence of an electric field. The charge carriers recombine in a forward-biased P-N junction as the electrons cross from the N-region and recombine with the holes existing in the P-region. Free electrons are in the conduction band of energy levels, while holes are in the valence energy band. Thus the energy level of the holes will be lesser than the energy levels of the electrons. Some portion of the energy must be dissipated in order to recombine the electrons and the holes. This energy is emitted in the form of heat and light.

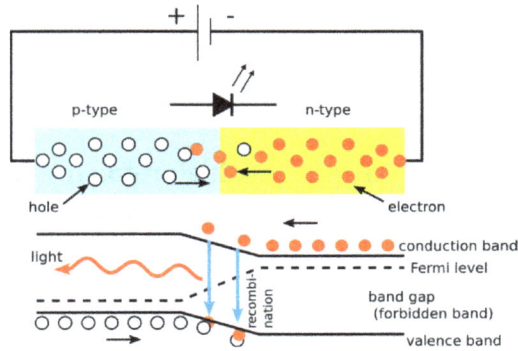

The inner workings of an LED, showing circuit (top) and band diagram (bottom)

The electrons dissipate energy in the form of heat for silicon and germanium diodes but in gallium arsenide phosphide (GaAsP) and gallium phosphide (GaP) semiconductors, the electrons dissipate energy by emitting photons. If the semiconductor is translucent, the junction becomes the source of light as it is emitted, thus becoming a light-emitting diode, but when the junction is reverse biased no light will be produced by the LED and, on the contrary, the device may also be damaged.

Technology

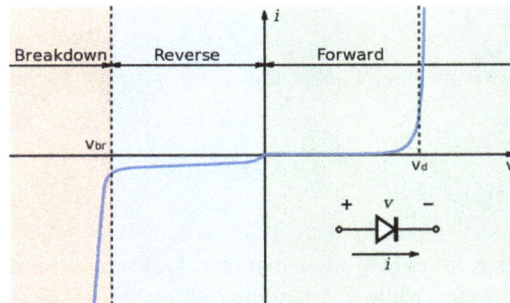

I-V diagram for a diode. An LED will begin to emit light when more than 2 or 3 volts is applied to it.

Physics

The LED consists of a chip of semiconducting material doped with impurities to create a *p-n junction*. As in other diodes, current flows easily from the p-side, or anode, to the n-side, or cathode, but not in the reverse direction. Charge-carriers—electrons and holes—flow into the junction from electrodes with different voltages. When an electron meets a hole, it falls into a lower energy level and releases energy in the form of a photon.

The wavelength of the light emitted, and thus its color, depends on the band gap energy of the materials forming the *p-n junction*. In silicon or germanium diodes, the electrons and holes usually recombine by a *non-radiative transition*, which produces no optical emission, because these are indirect band gap materials. The materials used for the LED have a direct band gap with energies corresponding to near-infrared, visible, or near-ultraviolet light.

LED development began with infrared and red devices made with gallium arsenide. Advances in materials science have enabled making devices with ever-shorter wavelengths, emitting light in a variety of colors.

LEDs are usually built on an n-type substrate, with an electrode attached to the p-type layer deposited on its surface. P-type substrates, while less common, occur as well. Many commercial LEDs, especially GaN/InGaN, also use sapphire substrate.

Most materials used for LED production have very high refractive indices. This means that much of the light will be reflected back into the material at the material/air surface interface. Thus, light extraction in LEDs is an important aspect of LED production, subject to much research and development.

Refractive Index

Bare uncoated semiconductors such as silicon exhibit a very high refractive index relative to open air, which prevents passage of photons arriving at sharp angles relative to the air-contacting surface of the semiconductor due to total internal reflection. This property affects both the light-emission efficiency of LEDs as well as the light-absorption efficiency of photovoltaic cells. The refractive index of silicon is 3.96 (at 590 nm), while air is 1.0002926.

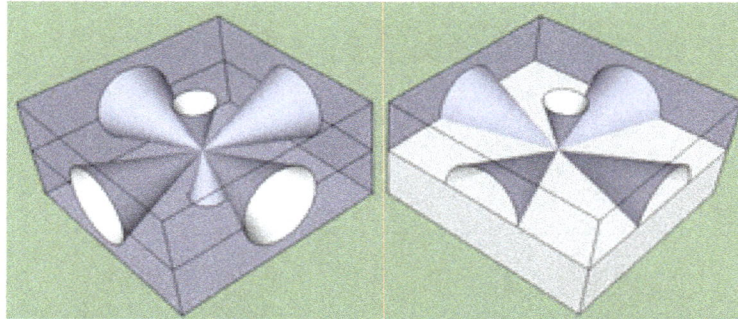

Idealized example of light emission cones in a semiconductor, for a single point-source emission zone. The left illustration is for a fully translucent wafer, while the right illustration shows the half-cones formed when the bottom layer is fully opaque. The light is actually emitted equally in all directions from the point-source, so the areas between the cones shows the large amount of trapped light energy that is wasted as heat.

In general, a flat-surface uncoated LED semiconductor chip will emit light only perpendicular to the semiconductor's surface, and a few degrees to the side, in a cone shape referred to as the *light cone, cone of light*, or the *escape cone*. The maximum angle of incidence is referred to as the critical angle. When this angle is exceeded, photons no longer escape the semiconductor but are instead reflected internally inside the semiconductor crystal as if it were a mirror.

Internal reflections can escape through other crystalline faces, if the incidence angle is low enough and the crystal is sufficiently transparent to not re-absorb the photon emission. But for a simple square LED with 90-degree angled surfaces on all sides, the faces all act as equal angle mirrors. In this case most of the light can not escape and is lost as waste heat in the crystal.

A convoluted chip surface with angled facets similar to a jewel or fresnel lens can increase light output by allowing light to be emitted perpendicular to the chip surface while far to the sides of the photon emission point.

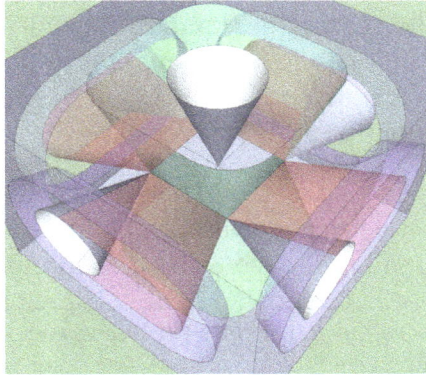

The light emission cones of a real LED wafer are far more complex than a single point-source light emission. The light emission zone is typically a two-dimensional plane between the wafers. Every atom across this plane has an individual set of emission cones. Drawing the billions of overlapping cones is impossible, so this is a simplified diagram showing the extents of all the emission cones combined. The larger side cones are clipped to show the interior features and reduce image complexity; they would extend to the opposite edges of the two-dimensional emission plane.

The ideal shape of a semiconductor with maximum light output would be a microsphere with the photon emission occurring at the exact center, with electrodes penetrating to the center to contact at the emission point. All light rays emanating from the center would be perpendicular to the entire surface of the sphere, resulting in no internal reflections. A hemispherical semiconductor would also work, with the flat back-surface serving as a mirror to back-scattered photons.

Transition Coatings

After the doping of the wafer, it is cut apart into individual dies. Each die is commonly called a chip.

Many LED semiconductor chips are encapsulated or potted in clear or colored molded plastic shells. The plastic shell has three purposes:

1. Mounting the semiconductor chip in devices is easier to accomplish.

2. The tiny fragile electrical wiring is physically supported and protected from damage.

3. The plastic acts as a refractive intermediary between the relatively high-index semiconductor and low-index open air.

The third feature helps to boost the light emission from the semiconductor by acting as a diffusing lens, allowing light to be emitted at a much higher angle of incidence from the light cone than the bare chip is able to emit alone.

Efficiency and Operational Parameters

Typical indicator LEDs are designed to operate with no more than 30–60 milliwatts (mW) of electrical power. Around 1999, Philips Lumileds introduced power LEDs capable of continuous use at one watt.

These LEDs used much larger semiconductor die sizes to handle the large power inputs. Also, the semi-conductor dies were mounted onto metal slugs to allow for heat removal from the LED die.

One of the key advantages of LED-based lighting sources is high luminous efficacy. White LEDs quickly matched and overtook the efficacy of standard incandescent lighting systems. In 2002, Lumileds made five-watt LEDs available with luminous efficacy of 18–22 lumens per watt (lm/W). For comparison, a conventional incandescent light bulb of 60–100 watts emits around 15 lm/W, and standard fluorescent lights emit up to 100 lm/W.

As of 2012, Philips had achieved the following efficacies for each color. The efficiency values show the physics - light power out per electrical power in. The lumen-per-watt efficacy value includes characteristics of the human eye, and is derived using the luminosity function.

Color	Wavelength range (nm)	Typical efficiency coefficient	Typical efficacy (lm/W)
Red	$620 < \lambda < 645$	0.39	72
Red-orange	$610 < \lambda < 620$	0.29	98
Green	$520 < \lambda < 550$	0.15	93
Cyan	$490 < \lambda < 520$	0.26	75
Blue	$460 < \lambda < 490$	0.35	37

In September 2003, a new type of blue LED was demonstrated by Cree that consumes 24 mW at 20 milliamperes (mA). This produced a commercially packaged white light giving 65 lm/W at 20 mA, becoming the brightest white LED commercially available at the time, and more than four times as efficient as standard incandescents. In 2006, they demonstrated a prototype with a record white LED luminous efficacy of 131 lm/W at 20 mA. Nichia Corporation has developed a white LED with luminous efficacy of 150 lm/W at a forward current of 20 mA. Cree's XLamp XM-L LEDs, commercially available in 2011, produce 100 lm/W at their full power of 10 W, and up to 160 lm/W at around 2 W input power. In 2012, Cree announced a white LED giving 254 lm/W, and 303 lm/W in March 2014. Practical general lighting needs high-power LEDs, of one watt or more. Typical operating currents for such devices begin at 350 mA.

These efficiencies are for the light-emitting diode only, held at low temperature in a lab. Since LEDs installed in real fixtures operate at higher temperature and with driver losses, real-world ef-ficiencies are much lower. United States Department of Energy (DOE) testing of commercial LED lamps designed to replace incandescent lamps or CFLs showed that average efficacy was still about 46 lm/W in 2009 (tested performance ranged from 17 lm/W to 79 lm/W).

Efficiency Droop

Efficiency droop is the decrease in luminous efficiency of LEDs as the electric current increases above tens of milliamperes.

This effect was initially theorized to be related to elevated temperatures. Scientists proved the opposite to be true that, although the life of an LED would be shortened, the efficiency droop is less severe at elevated temperatures. The mechanism causing efficiency droop was identified in 2007 as Auger recombination, which was taken with mixed reaction. In 2013, a study confirmed Auger recombination as the cause of efficiency droop.

In addition to being less efficient, operating LEDs at higher electric currents creates higher heat levels which compromise the lifetime of the LED. Because of this increased heating at higher currents, high-brightness LEDs have an industry standard of operating at only 350 mA, which is a compromise between light output, efficiency, and longevity.

Possible Solutions

Instead of increasing current levels, luminance is usually increased by combining multiple LEDs in one bulb. Solving the problem of efficiency droop would mean that household LED light bulbs would need fewer LEDs, which would significantly reduce costs.

Researchers at the U.S. Naval Research Laboratory have found a way to lessen the efficiency droop. They found that the droop arises from non-radiative Auger recombination of the injected carriers. They created quantum wells with a soft confinement potential to lessen the non-radiative Auger processes.

Researchers at Taiwan National Central University and Epistar Corp are developing a way to lessen the efficiency droop by using ceramic aluminium nitride (AlN) substrates, which are more thermally conductive than the commercially used sapphire. The higher thermal conductivity reduces self-heating effects.

Lifetime and Failure

Solid-state devices such as LEDs are subject to very limited wear and tear if operated at low currents and at low temperatures. Typical lifetimes quoted are 25,000 to 100,000 hours, but heat and current settings can extend or shorten this time significantly.

The most common symptom of LED (and diode laser) failure is the gradual lowering of light output and loss of efficiency. Sudden failures, although rare, can also occur. Early red LEDs were notable for their short service life. With the development of high-power LEDs the devices are subjected to higher junction temperatures and higher current densities than traditional devices. This causes stress on the material and may cause early light-output degradation. To quantitatively classify useful lifetime in a standardized manner it has been suggested to use L70 or L50, which are the runtimes (typically given in thousands of hours) at which a given LED reaches 70% and 50% of initial light output, respectively.

LED performance is temperature dependent. Most manufacturers' published ratings of LEDs are for an operating temperature of 25 °C (77 °F). LEDs used outdoors, such as traffic signals or in-pavement signal lights, and that are used in climates where the temperature within the light fixture gets very high, could result in low signal intensities or even failure.

Since LED efficacy is inversely proportional to operating temperature, LED technology is well suited for supermarket freezer lighting. Because LEDs produce less waste heat than incandescent lamps, their use in freezers can save on refrigeration costs as well. However, they may be more susceptible to frost and snow buildup than incandescent lamps, so some LED lighting systems have been designed with an added heating circuit. Additionally, research has developed heat sink technologies that will transfer heat produced within the junction to appropriate areas of the light fixture.

Colors and Materials

Conventional LEDs are made from a variety of inorganic semiconductor materials. The following table shows the available colors with wavelength range, voltage drop and material:

Color	Wavelength [nm]	Voltage drop [ΔV]	Semiconductor material
Infrared	$\lambda > 760$	$\Delta V < 1.63$	Gallium arsenide (GaAs) Aluminium gallium arsenide (AlGaAs)
Red	$610 < \lambda < 760$	$1.63 < \Delta V < 2.03$	Aluminium gallium arsenide (AlGaAs) Gallium arsenide phosphide (GaAsP) Aluminium gallium indium phosphide (AlGaInP) Gallium(III) phosphide (GaP)
Orange	$590 < \lambda < 610$	$2.03 < \Delta V < 2.10$	Gallium arsenide phosphide (GaAsP) Aluminium gallium indium phosphide (AlGaInP) Gallium(III) phosphide (GaP)
Yellow	$570 < \lambda < 590$	$2.10 < \Delta V < 2.18$	Gallium arsenide phosphide (GaAsP) Aluminium gallium indium phosphide (AlGaInP) Gallium(III) phosphide (GaP)
Green	$500 < \lambda < 570$	$1.9 < \Delta V < 4.0$	**Traditional green:** Gallium(III) phosphide (GaP) Aluminium gallium indium phosphide (AlGaInP) Aluminium gallium phosphide (AlGaP) **Pure green:** Indium gallium nitride (InGaN) / Gallium(III) nitride (GaN)
Blue	$450 < \lambda < 500$	$2.48 < \Delta V < 3.7$	Zinc selenide (ZnSe) Indium gallium nitride (InGaN) Silicon carbide (SiC) as substrate Silicon (Si) as substrate—under development
Violet	$400 < \lambda < 450$	$2.76 < \Delta V < 4.0$	Indium gallium nitride (InGaN)
Purple	Multiple types	$2.48 < \Delta V < 3.7$	Dual blue/red LEDs, blue with red phosphor, or white with purple plastic
Ultraviolet	$\lambda < 400$	$3 < \Delta V < 4.1$	Indium gallium nitride (InGaN) (385-400 nm) Diamond (235 nm) Boron nitride (215 nm) Aluminium nitride (AlN) (210 nm) Aluminium gallium nitride (AlGaN) Aluminium gallium indium nitride (AlGaInN)—down to 210 nm
Pink	Multiple types	$\Delta V \sim 3.3$	Blue with one or two phosphor layers, yellow with red, orange or pink phosphor added afterwards, white with pink plastic, or white phosphors with pink pigment or dye over top.
White	Broad spectrum	$2.8 < \Delta V < 4.2$	**Cool / Pure White:** Blue/UV diode with yellow phosphor **Warm White:** Blue diode with orange phosphor

Blue and Ultraviolet

The first blue-violet LED using magnesium-doped gallium nitride was made at Stanford University in 1972 by Herb Maruska and Wally Rhines, doctoral students in materials science and engineering. At the time Maruska was on leave from RCA Laboratories, where he collaborated with Jacques Pankove on related work. In 1971, the year after Maruska left for Stanford, his RCA colleagues Pankove and Ed Miller demonstrated the first blue electroluminescence from zinc-doped gallium nitride, though the subsequent device Pankove and Miller built, the first actual gallium nitride light-emitting diode, emitted green light. In 1974 the U.S. Patent Office awarded Maruska, Rhines and Stanford professor David Stevenson a patent for their work in 1972 (U.S. Patent US3819974 A) and today magnesium-doping of gallium nitride continues to be the basis for all commercial blue LEDs and laser diodes. These devices built in the early 1970s had too little light output to be of practical use and research into gallium nitride devices slowed. In August 1989, Cree introduced the first commercially available blue LED based on the indirect bandgap semiconductor, silicon carbide (SiC). SiC LEDs had very low efficiency, no more than about 0.03%, but did emit in the blue portion of the visible light spectrum.

Blue LEDs

In the late 1980s, key breakthroughs in GaN epitaxial growth and p-type doping ushered in the modern era of GaN-based optoelectronic devices. Building upon this foundation, Dr. Moustakas at Boston University patented a method for producing high-brightness blue LEDs using a new two-step process. Two years later, in 1993, high-brightness blue LEDs were demonstrated again by Shuji Nakamura of Nichia Corporation using a gallium nitride growth process similar to Dr. Moustakas's. Both Dr. Moustakas and Mr. Nakamura were issued separate patents, which confused the issue of who was the original inventor (partly because although Dr. Moustakas invented his first, Dr. Nakamura filed first).This new development revolutionized LED lighting, making high-power blue light sources practical, leading to the development of technologies like Blu-ray, as well as allowing the bright high resolution screens of modern tablets and phones.

Nakamura was awarded the 2006 Millennium Technology Prize for his invention. Nakamura, Hiroshi Amano and Isamu Akasaki were awarded the Nobel Prize in Physics in 2014 for the invention of the blue LED. In 2015, a US court ruled that three companies (i.e. the litigants who had not previously settled out of court) that had licensed Mr. Nakamura's patents for production in the United States had infringed Dr. Moustakas's prior patent, and order them to pay licensing fees of not less than 13 million USD.

External Video

"The Original Blue LED", Chemical Heritage Foundation

By the late 1990s, blue LEDs became widely available. They have an active region consisting of one or more InGaN quantum wells sandwiched between thicker layers of GaN, called cladding layers. By varying the relative In/Ga fraction in the InGaN quantum wells, the light emission can in theory be varied from violet to amber. Aluminium gallium nitride (AlGaN) of varying Al/Ga fraction can be used to manufacture the cladding and quantum well layers for ultraviolet LEDs, but these devices have not yet reached the level of efficiency and technological maturity of InGaN/GaN blue/green devices. If un-alloyed GaN is used in this case to form the active quantum well layers, the device will emit near-ultraviolet light with a peak wavelength centred around 365 nm. Green LEDs manufactured from the InGaN/GaN system are far more efficient and brighter than green LEDs produced with non-nitride material systems, but practical devices still exhibit efficiency too low for high-brightness applications.

With nitrides containing aluminium, most often AlGaN and AlGaInN, even shorter wavelengths are achievable. Ultraviolet LEDs in a range of wavelengths are becoming available on the market. Near-UV emitters at wavelengths around 375–395 nm are already cheap and often encountered, for example, as black light lamp replacements for inspection of anti-counterfeiting UV watermarks in some documents and paper currencies. Shorter-wavelength diodes, while substantially more expensive, are commercially available for wavelengths down to 240 nm. As the photosensitivity of microorganisms approximately matches the absorption spectrum of DNA, with a peak at about 260 nm, UV LED emitting at 250–270 nm are to be expected in prospective disinfection and sterilization devices. Recent research has shown that commercially available UVA LEDs (365 nm) are already effective disinfection and sterilization devices. UV-C wavelengths were obtained in laboratories using aluminium nitride (210 nm), boron nitride (215 nm) and diamond (235 nm).

RGB

RGB LEDs consist of one red, one green, and one blue LED. By independently attenuating each of the three, RGB LEDs are capable of producing a wide color gamut.

RGB LED

White

There are two primary ways of producing white light-emitting diodes (WLEDs), LEDs that generate high-intensity white light. One is to use individual LEDs that emit three primary colors—red, green, and blue—and then mix all the colors to form white light. The other is to use a phosphor material to convert monochromatic light from a blue or UV LED to broad-spectrum white light, much in the same way a fluorescent light bulb works.

There are three main methods of mixing colors to produce white light from an LED:

- blue LED + green LED + red LED (color mixing; can be used as backlighting for displays)

- near-UV or UV LED + RGB phosphor (an LED producing light with a wavelength shorter than blue's is used to excite an RGB phosphor)

- blue LED + yellow phosphor (two complementary colors combine to form white light; more efficient than first two methods and more commonly used)

Because of metamerism, it is possible to have quite different spectra that appear white. However, the appearance of objects illuminated by that light may vary as the spectrum varies.

RGB Systems

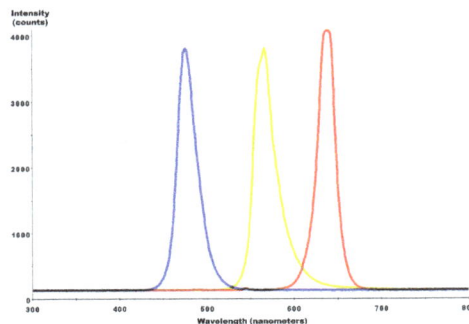

Combined spectral curves for blue, yellow-green, and high-brightness red solid-state semiconductor LEDs. FWHM spectral bandwidth is approximately 24–27 nm for all three colors.

White light can be formed by mixing differently colored lights; the most common method is to use red, green, and blue (RGB). Hence the method is called multi-color white LEDs (sometimes referred to as RGB LEDs). Because these need electronic circuits to control the blending and diffusion of different colors, and because the individual color LEDs typically have slightly different emission patterns (leading to variation of the color depending on direction) even if they are made as a single unit, these are seldom used to produce white lighting. Nonetheless, this method has many applications because of the flexibility of mixing different colors, and in principle, this mechanism also has higher quantum efficiency in producing white light.

RGB LED

There are several types of multi-color white LEDs: di-, tri-, and tetrachromatic white LEDs. Several key factors that play among these different methods, include color stability, color rendering capability, and luminous efficacy. Often, higher efficiency will mean lower color rendering, presenting a trade-off between the luminous efficacy and color rendering. For example, the dichromatic white LEDs have the best luminous efficacy (120 lm/W), but the lowest color rendering capability. However, although tetrachromatic white LEDs have excellent color rendering capability, they often have poor luminous efficacy. Trichromatic white LEDs are in between, having both good luminous efficacy (>70 lm/W) and fair color rendering capability.

One of the challenges is the development of more efficient green LEDs. The theoretical maximum for green LEDs is 683 lumens per watt but as of 2010 few green LEDs exceed even 100 lumens per watt. The blue and red LEDs get closer to their theoretical limits.

Multi-color LEDs offer not merely another means to form white light but a new means to form light of different colors. Most perceivable colors can be formed by mixing different amounts of three primary colors. This allows precise dynamic color control. As more effort is devoted to investigating this method, multi-color LEDs should have profound influence on the fundamental method that we use to produce and control light color. However, before this type of LED can play a role on the market, several technical problems must be solved. These include that this type of LED's emission power decays exponentially with rising temperature, resulting in a substantial change in color stability. Such problems inhibit and may preclude industrial use. Thus, many new package designs aimed at solving this problem have been proposed and their results are now being reproduced by researchers and scientists.

Correlated color temperature (CCT) dimming for LED technology is regarded as a difficult task, since binning, age and temperature drift effects of LEDs change the actual color value output. Feedback loop systems are used for example with color sensors, to actively monitor and control the color output of multiple color mixing LEDs.

Phosphor-based LEDs

This method involves coating LEDs of one color (mostly blue LEDs made of InGaN) with phosphors of different colors to form white light; the resultant LEDs are called phosphor-based or phosphor-converted white LEDs (pcLEDs). A fraction of the blue light undergoes the Stokes shift being transformed from shorter wavelengths to longer. Depending on the color of the original LED, phosphors of different colors can be employed. If several phosphor layers of distinct colors are applied, the emitted spectrum is broadened, effectively raising the color rendering index (CRI) value of a given LED.

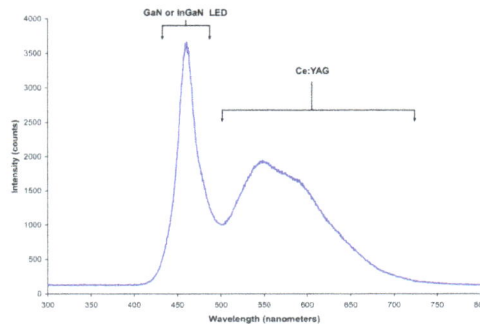

Spectrum of a white LED showing blue light directly emitted by the GaN-based LED (peak at about 465 nm) and the more broadband Stokes-shifted light emitted by the Ce3+:YAG phosphor, which emits at roughly 500–700 nm

Phosphor-based LED efficiency losses are due to the heat loss from the Stokes shift and also other phosphor-related degradation issues. Their luminous efficacies compared to normal LEDs depend on the spectral distribution of the resultant light output and the original wavelength of the LED itself. For example, the luminous efficacy of a typical YAG yellow phosphor based white LED ranges from 3 to 5 times the luminous efficacy of the original blue LED because of the human eye's greater sensitivity to yellow than to blue (as modeled in the luminosity function). Due to the simplicity of manufacturing the phosphor method is still the most popular method for making high-intensity white LEDs. The design and production of a light source or light fixture using a monochrome emitter with phosphor conversion is simpler and cheaper than a complex RGB system, and the majority of high-intensity white LEDs presently on the market are manufactured using phosphor light conversion.

Among the challenges being faced to improve the efficiency of LED-based white light sources is the development of more efficient phosphors. As of 2010, the most efficient yellow phosphor is still the YAG phosphor, with less than 10% Stoke shift loss. Losses attributable to internal optical losses due to re-absorption in the LED chip and in the LED packaging itself account typically for another 10% to 30% of efficiency loss. Currently, in the area of phosphor LED development, much effort is being spent on optimizing these devices to higher light output and higher operation temperatures. For instance, the efficiency can be raised by adapting better package design or by using a more suitable type of phosphor. Conformal coating process is frequently used to address the issue of varying phosphor thickness.

Some phosphor-based white LEDs encapsulate InGaN blue LEDs inside phosphor-coated epoxy. Alternatively, the LED might be paired with a remote phosphor, a preformed polycarbonate piece coated with the phosphor material. Remote phosphors provide more diffuse light, which is desirable for many applications. Remote phosphor designs are also more tolerant of variations in the LED emissions spectrum. A common yellow phosphor material is cerium-doped yttrium aluminium garnet (Ce^{3+}:YAG).

White LEDs can also be made by coating near-ultraviolet (NUV) LEDs with a mixture of high-efficiency europium-based phosphors that emit red and blue, plus copper and aluminium-doped zinc sulfide (ZnS:Cu, Al) that emits green. This is a method analogous to the way fluorescent lamps work. This method is less efficient than blue LEDs with YAG:Ce phosphor, as the Stokes shift is larger, so more energy is converted to heat, but yields light with better spectral characteristics, which render color better. Due to the higher radiative output of the ultraviolet LEDs than of the blue ones, both methods offer comparable brightness. A concern is that UV light may leak from a malfunctioning light source and cause harm to human eyes or skin.

Other White LEDs

Another method used to produce experimental white light LEDs used no phosphors at all and was based on homoepitaxially grown zinc selenide (ZnSe) on a ZnSe substrate that simultaneously emitted blue light from its active region and yellow light from the substrate.

A new style of wafers composed of gallium-nitride-on-silicon (GaN-on-Si) is being used to produce white LEDs using 200-mm silicon wafers. This avoids the typical costly sapphire substrate in relatively small 100- or 150-mm wafer sizes. The sapphire apparatus must be coupled with a mirror-like collector to reflect light that would otherwise be wasted. It is predicted that by 2020, 40% of all GaN LEDs will be made with GaN-on-Si. Manufacturing large sapphire material is difficult, while large silicon material is cheaper and more abundant. LED companies shifting from using sapphire to silicon should be a minimal investment.

Organic Light-emitting Diodes (OLEDs)

In an organic light-emitting diode (OLED), the electroluminescent material comprising the emissive layer of the diode is an organic compound. The organic material is electrically conductive due to the delocalization of pi electrons caused by conjugation over all or part of the molecule, and the material therefore functions as an organic semiconductor. The organic materials can be small organic molecules in a crystalline phase, or polymers.

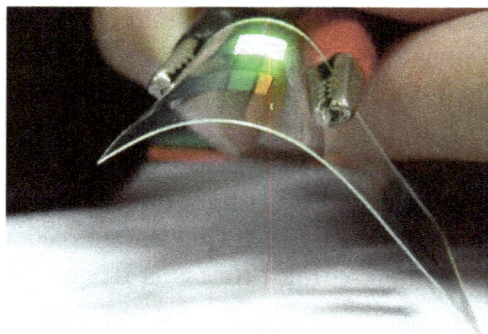

Demonstration of a flexible OLED device

The potential advantages of OLEDs include thin, low-cost displays with a low driving voltage, wide viewing angle, and high contrast and color gamut. Polymer LEDs have the added benefit of printable and flexible displays. OLEDs have been used to make visual displays for portable electronic devices such as cellphones, digital cameras, and MP3 players while possible future uses include lighting and televisions.

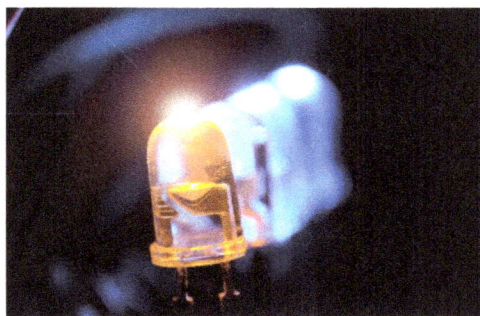

Orange light-emitting diode

Quantum Dot LEDs

Quantum dots (QD) are semiconductor nanocrystals that possess unique optical properties. Their emission color can be tuned from the visible throughout the infrared spectrum. This allows quantum dot LEDs to create almost any color on the CIE diagram. This provides more color options and better color rendering than white LEDs since the emission spectrum is much narrower, characteristic of quantum confined states. There are two types of schemes for QD excitation. One uses photo excitation with a primary light source LED (typically blue or UV LEDs are used). The other is direct electrical excitation first demonstrated by Alivisatos et al.

One example of the photo-excitation scheme is a method developed by Michael Bowers, at Vanderbilt University in Nashville, involving coating a blue LED with quantum dots that glow white in response to the blue light from the LED. This method emits a warm, yellowish-white light similar to that made by incandescent light bulbs. Quantum dots are also being considered for use in white light-emitting diodes in liquid crystal display (LCD) televisions.

In February 2011 scientists at PlasmaChem GmbH were able to synthesize quantum dots for LED applications and build a light converter on their basis, which was able to efficiently convert light from blue to any other color for many hundred hours. Such QDs can be used to emit visible or near infrared light of any wavelength being excited by light with a shorter wavelength.

The structure of QD-LEDs used for the electrical-excitation scheme is similar to basic design of OLEDs. A layer of quantum dots is sandwiched between layers of electron-transporting and hole-transporting materials. An applied electric field causes electrons and holes to move into the quantum dot layer and recombine forming an exciton that excites a QD. This scheme is commonly studied for quantum dot display. The tunability of emission wavelengths and narrow bandwidth is also beneficial as excitation sources for fluorescence imaging. Fluorescence near-field scanning optical microscopy (NSOM) utilizing an integrated QD-LED has been demonstrated.

In February 2008, a luminous efficacy of 300 lumens of visible light per watt of radiation (not per electrical watt) and warm-light emission was achieved by using nanocrystals.

Types

The main types of LEDs are miniature, high-power devices and custom designs such as alphanumeric or multi-color.

LEDs are produced in a variety of shapes and sizes. The color of the plastic lens is often the same as the actual color of light emitted, but not always. For instance, purple plastic is often used for infrared LEDs, and most blue devices have colorless housings. Modern high-power LEDs such as those used for lighting and backlighting are generally found in surface-mount technology (SMT) packages (not shown).

Miniature

These are mostly single-die LEDs used as indicators, and they come in various sizes from 2 mm to 8 mm, through-hole and surface mount packages. They usually do not use a separate heat sink. Typical current ratings ranges from around 1 mA to above 20 mA. The small size sets a natural upper boundary on power consumption due to heat caused by the high current density and need for a heat sink.

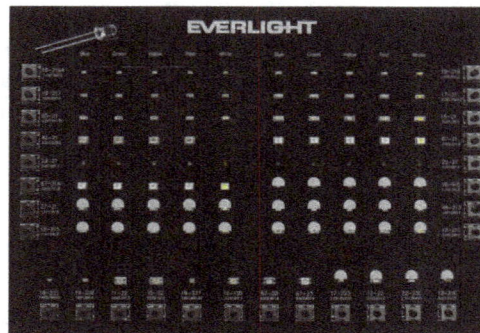

Photo of miniature surface mount LEDs in most common sizes. They can be much smaller than a traditional 5 mm lamp type LED which is shown on the upper left corner.

Very small (1.6x1.6x0.35 mm) red, green, and blue surface mount miniature LED package with gold wire bonding details.

Common package shapes include round, with a domed or flat top, rectangular with a flat top (as used in bar-graph displays), and triangular or square with a flat top. The encapsulation may also be clear or tinted to improve contrast and viewing angle.

Researchers at the University of Washington have invented the thinnest LED. It is made of two-dimensional (2-D) flexible materials. It is three atoms thick, which is 10 to 20 times thinner than three-dimensional (3-D) LEDs and is also 10,000 times smaller than the thickness of a human hair. These 2-D LEDs are going to make it possible to create smaller, more energy-efficient lighting, optical communication and nano lasers.

There are three main categories of miniature single die LEDs:

Low-current

> Typically rated for 2 mA at around 2 V (approximately 4 mW consumption)

Standard

> 20 mA LEDs (ranging from approximately 40 mW to 90 mW) at around:

- 1.9 to 2.1 V for red, orange, yellow, and traditional green

- 3.0 to 3.4 V for pure green and blue

- 2.9 to 4.2 V for violet, pink, purple and white

Ultra-high-output

> 20 mA at approximately 2 or 4–5 V, designed for viewing in direct sunlight

5 V and 12 V LEDs are ordinary miniature LEDs that incorporate a suitable series resistor for direct connection to a 5 V or 12 V supply.

High-power

High-power LEDs (HP-LEDs) or high-output LEDs (HO-LEDs) can be driven at currents from hundreds of mA to more than an ampere, compared with the tens of mA for other LEDs. Some can emit over a thousand lumens. LED power densities up to 300 W/cm^2 have been achieved. Since overheating is destructive, the HP-LEDs must be mounted on a heat sink to allow for heat dissipation. If the heat from a HP-LED is not removed, the device will fail in seconds. One HP-LED can often replace an incandescent bulb in a flashlight, or be set in an array to form a powerful LED lamp.

High-power light-emitting diodes attached to an LED star base (Luxeon, Lumileds)

Some well-known HP-LEDs in this category are the Nichia 19 series, Lumileds Rebel Led, Osram Opto Semiconductors Golden Dragon, and Cree X-lamp. As of September 2009, some HP-LEDs manufactured by Cree now exceed 105 lm/W.

Examples for Haitz's law, which predicts an exponential rise in light output and efficacy of LEDs over time, are the CREE XP-G series LED which achieved 105 lm/W in 2009 and the Nichia 19 series with a typical efficacy of 140 lm/W, released in 2010.

AC Driven

LEDs have been developed by Seoul Semiconductor that can operate on AC power without the need for a DC converter. For each half-cycle, part of the LED emits light and part is dark, and this is reversed during the next half-cycle. The efficacy of this type of HP-LED is typically 40 lm/W. A large number of LED elements in series may be able to operate directly from line voltage. In 2009, Seoul Semiconductor released a high DC voltage LED, named as 'Acrich MJT', capable of being driven from AC power with a simple controlling circuit. The low-power dissipation of these LEDs affords them more flexibility than the original AC LED design.

Application-specific Variations

Flashing

Flashing LEDs are used as attention seeking indicators without requiring external electronics. Flashing LEDs resemble standard LEDs but they contain an integrated multivibrator circuit that causes the LED to flash with a typical period of one second. In diffused lens LEDs, this circuit is visible as a small black dot. Most flashing LEDs emit light of one color, but more sophisticated devices can flash between multiple colors and even fade through a color sequence using RGB color mixing.

Bi-color

Bi-color LEDs contain two different LED emitters in one case. There are two types of these. One type consists of two dies connected to the same two leads antiparallel to each other. Current flow in one direction emits one color, and current in the opposite direction emits the other color. The other type consists of two dies with separate leads for both dies and another lead for common anode or cathode, so that they can be controlled independently.

Tri-color

Tri-color LEDs contain three different LED emitters in one case. Each emitter is connected to a separate lead so they can be controlled independently. A four-lead arrangement is typical with one common lead (anode or cathode) and an additional lead for each color.

RGB

RGB LEDs are tri-color LEDs with red, green, and blue emitters, in general using a four-wire connection with one common lead (anode or cathode). These LEDs can have either common positive or common negative leads. Others however, have only two leads (positive and negative) and have a built-in tiny electronic control unit.

Decorative-multicolor

Decorative-multicolor LEDs incorporate several emitters of different colors supplied by only two lead-out wires. Colors are switched internally by varying the supply voltage.

Alphanumeric

Alphanumeric LEDs are available in seven-segment, starburst and dot-matrix format. Seven-segment displays handle all numbers and a limited set of letters. Starburst displays can display all letters. Dot-matrix displays typically use 5x7 pixels per character. Seven-segment LED displays were in widespread use in the 1970s and 1980s, but rising use of liquid crystal displays, with their lower power needs and greater display flexibility, has reduced the popularity of numeric and alphanumeric LED displays.

Digital-RGB

Digital-RGB LEDs are RGB LEDs that contain their own "smart" control electronics. In addition to power and ground, these provide connections for data-in, data-out, and sometimes a clock or strobe signal. These are connected in a daisy chain, with the data in of the first LED sourced by a microprocessor, which can control the brightness and color of each LED independently of the others. They are used where a combination of maximum control and minimum visible electronics are needed such as strings for Christmas and LED matrices. Some even have refresh rates in the kHz range, allowing for basic video applications.

Filament

An LED filament consists of multiple LED dice connected in series on a common longitudinal substrate that form a thin rod reminiscent of a traditional incandescent filament. These are being used as a low cost decorative alternative for traditional light bulbs that are being phased out in many countries. The filaments require a rather high voltage to light to nominal brightness, allowing them to work efficiently and simply with mains voltages. Often a simple rectifier and capacitive current limiting are employed to create a low-cost replacement for a traditional light bulb without the complexity of creating a low voltage, high current converter which is required by single die LEDs. Usually they are packaged in a sealed enclosure with a shape similar to lamps they were designed to replace (e.g. a bulb), and filled with inert nitrogen or carbon dioxide gas to remove heat efficiently.

Considerations for Use

Power Sources

The current–voltage characteristic of an LED is similar to other diodes, in that the current is dependent exponentially on the voltage (Shockley diode equation). This means that a small change in voltage can cause a large change in current. If the applied voltage exceeds the LED's forward voltage drop by a small amount, the current rating may be exceeded by a large amount, potentially damaging or destroying the LED. The typical solution is to use constant-current power supplies to keep the current below the LED's maximum current rating. Since most common power sources (batteries, mains) are constant-voltage sources, most LED fixtures must include a power converter, at least a current-limiting resistor. However, the high resistance of three-volt coin cells combined with the high differential resistance of nitride-based LEDs makes it possible to power such an LED from such a coin cell without an external resistor.

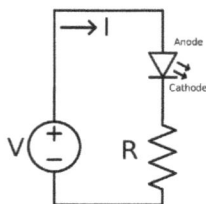

Simple LED circuit with resistor for current limiting

Electrical Polarity

As with all diodes, current flows easily from p-type to n-type material. However, no current flows and no light is emitted if a small voltage is applied in the reverse direction. If the reverse voltage grows large enough to exceed the breakdown voltage, a large current flows and the LED may be damaged. If the reverse current is sufficiently limited to avoid damage, the reverse-conducting LED is a useful noise diode.

Safety and Health

The vast majority of devices containing LEDs are "safe under all conditions of normal use", and so are classified as "Class 1 LED product"/"LED Klasse 1". At present, only a few LEDs—extremely bright LEDs that also have a tightly focused viewing angle of 8° or less—could, in theory, cause temporary blindness, and so are classified as "Class 2". The opinion of the French Agency for Food, Environmental and Occupational Health & Safety (ANSES) of 2010, on the health issues concerning LEDs, suggested banning public use of lamps which were in the moderate Risk Group 2, especially those with a high blue component in places frequented by children. In general, laser safety regulations—and the "Class 1", "Class 2", etc. system—also apply to LEDs.

While LEDs have the advantage over fluorescent lamps that they do not contain mercury, they may contain other hazardous metals such as lead and arsenic. Regarding the toxicity of LEDs when treated as waste, a study published in 2011 stated: "According to federal standards, LEDs are not hazardous except for low-intensity red LEDs, which leached Pb [lead] at levels exceeding regulatory limits (186 mg/L; regulatory limit: 5). However, according to California regulations, excessive levels of copper (up to 3892 mg/kg; limit: 2500), lead (up to 8103 mg/kg; limit: 1000), nickel (up to 4797 mg/kg; limit: 2000), or silver (up to 721 mg/kg; limit: 500) render all except low-intensity yellow LEDs hazardous."

Advantages

- Efficiency: LEDs emit more lumens per watt than incandescent light bulbs. The efficiency of LED lighting fixtures is not affected by shape and size unlike fluorescent light bulbs or tubes.

- Color: LEDs can emit light of an intended color without using any color filters as traditional lighting methods need. This is more efficient and can lower initial costs.

- Size: LEDs can be very small (smaller than 2 mm²) and are easily attached to printed circuit boards.

- Warmup time: LEDs light up very quickly. A typical red indicator LED will achieve full brightness in under a microsecond. LEDs used in communications devices can have even faster response times.

- Cycling: LEDs are ideal for uses subject to frequent on-off cycling, unlike incandescent and fluorescent lamps that fail faster when cycled often, or high-intensity discharge lamps (HID lamps) that require a long time before restarting.

- Dimming: LEDs can very easily be dimmed either by pulse-width modulation or lowering the forward current. This pulse-width modulation is why LED lights, particularly head-lights on cars, when viewed on camera or by some people, appear to be flashing or flickering. This is a type of stroboscopic effect.

- Cool light: In contrast to most light sources, LEDs radiate very little heat in the form of IR that can cause damage to sensitive objects or fabrics. Wasted energy is dispersed as heat through the base of the LED.

- Slow failure: LEDs mostly fail by dimming over time, rather than the abrupt failure of incandescent bulbs.

- Lifetime: LEDs can have a relatively long useful life. One report estimates 35,000 to 50,000 hours of useful life, though time to complete failure may be longer. Fluorescent tubes typically are rated at about 10,000 to 15,000 hours, depending partly on the conditions of use, and incandescent light bulbs at 1,000 to 2,000 hours. Several DOE demonstrations have shown that reduced maintenance costs from this extended lifetime, rather than energy savings, is the primary factor in determining the payback period for an LED product.

- Shock resistance: LEDs, being solid-state components, are difficult to damage with external shock, unlike fluorescent and incandescent bulbs, which are fragile.

- Focus: The solid package of the LED can be designed to focus its light. Incandescent and fluorescent sources often require an external reflector to collect light and direct it in a usable manner. For larger LED packages total internal reflection (TIR) lenses are often used to the same effect. However, when large quantities of light are needed many light sources are usually deployed, which are difficult to focus or collimate towards the same target.

Disadvantages

- Initial price: LEDs are currently slightly more expensive (price per lumen) on an initial capital cost basis, than other lighting technologies. As of March 2014, at least one manufacturer claims to have reached $1 per kilolumen. The additional expense partially stems from the relatively low lumen output and the drive circuitry and power supplies needed.

- Temperature dependence: LED performance largely depends on the ambient temperature of the operating environment – or "thermal management" properties. Over-driving an LED in high ambient temperatures may result in overheating the LED package, eventually leading to device failure. An adequate heat sink is needed to maintain long life. This is especially important in automotive, medical, and military uses where devices must operate over a wide range of temperatures, which require low failure rates. Toshiba has produced LEDs with an operating temperature range of -40 to 100 °C, which suits the LEDs for both indoor and outdoor use in applications such as lamps, ceiling lighting, street lights, and floodlights.

- Voltage sensitivity: LEDs must be supplied with a voltage above their threshold voltage and a current below their rating. Current and lifetime change greatly with a small change in applied voltage. They thus require a current-regulated supply (usually just a series resistor for indicator LEDs).

- Color rendition: Most cool-white LEDs have spectra that differ significantly from a black body radiator like the sun or an incandescent light. The spike at 460 nm and dip at 500 nm can cause the color of objects to be perceived differently under cool-white LED illumination than sunlight or incandescent sources, due to metamerism, red surfaces being rendered particularly poorly by typical phosphor-based cool-white LEDs.

- Area light source: Single LEDs do not approximate a point source of light giving a spherical light distribution, but rather a lambertian distribution. So LEDs are difficult to apply to uses needing a spherical light field; however, different fields of light can be manipulated by the application of different optics or "lenses". LEDs cannot provide divergence below a few degrees. In contrast, lasers can emit beams with divergences of 0.2 degrees or less.

- Electrical polarity: Unlike incandescent light bulbs, which illuminate regardless of the electrical polarity, LEDs will only light with correct electrical polarity. To automatically match source polarity to LED devices, rectifiers can be used.

- Blue hazard: There is a concern that blue LEDs and cool-white LEDs are now capable of exceeding safe limits of the so-called blue-light hazard as defined in eye safety specifications such as ANSI/IESNA RP-27.1–05: Recommended Practice for Photobiological Safety for Lamp and Lamp Systems.

- Light pollution: Because white LEDs, especially those with high color temperature, emit much more short wavelength light than conventional outdoor light sources such as high-pressure sodium vapor lamps, the increased blue and green sensitivity of scotopic vision means that white LEDs used in outdoor lighting cause substantially more sky glow. The American Medical Association warned on the use of high blue content white LEDs in street lighting, due to their higher impact on human health and environment, compared to low blue content light sources (e.g. High Pressure Sodium, PC amber LEDs, and low CCT LEDs).

- Efficiency droop: The efficiency of LEDs decreases as the electric current increases. Heating also increases with higher currents which compromises the lifetime of the LED. These effects put practical limits on the current through an LED in high power applications.

- Impact on insects: LEDs are much more attractive to insects than sodium-vapor lights, so much so that there has been speculative concern about the possibility of disruption to food webs.

- Use in winter conditions: Since they do not give off much heat in comparison to traditional electrical lights, LED lights used for traffic control can have snow obscuring them, leading to accidents.

Applications

LED uses fall into four major categories:

- Visual signals where light goes more or less directly from the source to the human eye, to convey a message or meaning

- Illumination where light is reflected from objects to give visual response of these objects

- Measuring and interacting with processes involving no human vision

- Narrow band light sensors where LEDs operate in a reverse-bias mode and respond to incident light, instead of emitting light

Indicators and Signs

The low energy consumption, low maintenance and small size of LEDs has led to uses as status indicators and displays on a variety of equipment and installations. Large-area LED displays are used as stadium displays and as dynamic decorative displays. Thin, lightweight message displays are used at airports and railway stations, and as destination displays for trains, buses, trams, and ferries.

Red and green LED traffic signals

One-color light is well suited for traffic lights and signals, exit signs, emergency vehicle lighting, ships' navigation lights or lanterns (chromacity and luminance standards being set under the Convention on the International Regulations for Preventing Collisions at Sea 1972, Annex I and the CIE) and LED-based Christmas lights. In cold climates, LED traffic lights may remain snow-covered. Red or yellow LEDs are used in indicator and alphanumeric displays in environments where night vision must be retained: aircraft cockpits, submarine and ship bridges, astronomy observatories, and in the field, e.g. night time animal watching and military field use.

Because of their long life, fast switching times, and their ability to be seen in broad daylight due to their high output and focus, LEDs have been used in brake lights for cars' high-mounted brake lights, trucks, and buses, and in turn signals for some time, but many vehicles now use LEDs for their rear light clusters. The use in brakes improves safety, due to a great reduction in the time needed to light fully, or faster rise time, up to 0.5 second faster than an incandescent bulb. This gives drivers behind more time to react. In a dual intensity circuit (rear markers and brakes) if the LEDs are not pulsed at a fast enough frequency, they can create a phantom array, where ghost images of the LED will appear if the eyes quickly scan across the array. White LED headlamps are starting to be used. Using LEDs has styling advantages because LEDs can form much thinner lights than incandescent lamps with parabolic reflectors.

Automotive applications for LEDs continue to grow.

Due to the relative cheapness of low output LEDs, they are also used in many temporary uses such as glowsticks, throwies, and the photonic textile Lumalive. Artists have also used LEDs for LED art.

Weather and all-hazards radio receivers with Specific Area Message Encoding (SAME) have three LEDs: red for warnings, orange for watches, and yellow for advisories and statements whenever issued.

Lighting

With the development of high-efficiency and high-power LEDs, it has become possible to use LEDs in lighting and illumination. To encourage the shift to LED lamps and other high-efficiency lighting, the US Department of Energy has created the L Prize competition. The Philips Lighting North America LED bulb won the first competition on August 3, 2011 after successfully completing 18 months of intensive field, lab, and product testing.

LEDs are used as street lights and in other architectural lighting. The mechanical robustness and long lifetime is used in automotive lighting on cars, motorcycles, and bicycle lights. LED light emission may be efficiently controlled by using nonimaging optics principles.

LED street lights are employed on poles and in parking garages. In 2007, the Italian village of Torraca was the first place to convert its entire illumination system to LEDs.

LEDs are used in aviation lighting. Airbus has used LED lighting in its Airbus A320 Enhanced since 2007, and Boeing uses LED lighting in the 787. LEDs are also being used now in airport and heliport lighting. LED airport fixtures currently include medium-intensity runway lights, runway centerline lights, taxiway centerline and edge lights, guidance signs, and obstruction lighting.

LEDs are also used as a light source for DLP projectors, and to backlight LCD televisions (referred to as LED TVs) and laptop displays. RGB LEDs raise the color gamut by as much as 45%. Screens for TV and computer displays can be made thinner using LEDs for backlighting.

The lack of IR or heat radiation makes LEDs ideal for stage lights using banks of RGB LEDs that can easily change color and decrease heating from traditional stage lighting, as well as medical lighting where IR-radiation can be harmful. In energy conservation, the lower heat output of LEDs also means air conditioning (cooling) systems have less heat in need of disposal.

LEDs are small, durable and need little power, so they are used in handheld devices such as flashlights. LED strobe lights or camera flashes operate at a safe, low voltage, instead of the 250+ volts commonly found in xenon flashlamp-based lighting. This is especially useful in cameras on mobile phones, where space is at a premium and bulky voltage-raising circuitry is undesirable.

LEDs are used for infrared illumination in night vision uses including security cameras. A ring of LEDs around a video camera, aimed forward into a retroreflective background, allows chroma keying in video productions.

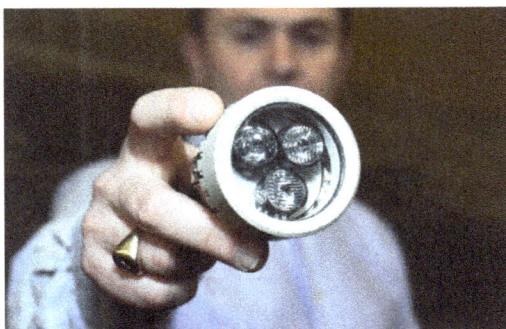

LED to be used for miners, to increase visibility inside mines

LEDs are used in mining operations, as cap lamps to provide light for miners. Research has been done to improve LEDs for mining, to reduce glare and to increase illumination, reducing risk of injury to the miners.

LEDs are now used commonly in all market areas from commercial to home use: standard lighting, AV, stage, theatrical, architectural, and public installations, and wherever artificial light is used.

LEDs are increasingly finding uses in medical and educational applications, for example as mood enhancement,and new technologies such as AmBX, exploiting LED versatility. NASA has even sponsored research for the use of LEDs to promote health for astronauts.

Data Communication and Other Signalling

Light can be used to transmit data and analog signals. For example, lighting white LEDs can be used in systems assisting people to navigate in closed spaces while searching necessary rooms or objects.

Assistive listening devices in many theaters and similar spaces use arrays of infrared LEDs to send sound to listeners' receivers. Light-emitting diodes (as well as semiconductor lasers) are used to send data over many types of fiber optic cable, from digital audio over TOSLINK cables to the very high bandwidth fiber links that form the Internet backbone. For some time, computers were commonly equipped with IrDA interfaces, which allowed them to send and receive data to nearby machines via infrared.

Because LEDs can cycle on and off millions of times per second, very high data bandwidth can be achieved.

Sustainable Lighting

Efficient lighting is needed for sustainable architecture. In 2009, US Department of Energy testing results on LED lamps showed an average efficacy of 35 lm/W, below that of typical CFLs, and as low as 9 lm/W, worse than standard incandescent bulbs. A typical 13-watt LED lamp emitted 450 to 650 lumens, which is equivalent to a standard 40-watt incandescent bulb.

However, as of 2011, there are LED bulbs available as efficient as 150 lm/W and even inexpensive low-end models typically exceed 50 lm/W, so that a 6-watt LED could achieve the same results as a standard 40-watt incandescent bulb. The latter has an expected lifespan of 1,000 hours, whereas an LED can continue to operate with reduced efficiency for more than 50,000 hours.

See the chart below for a comparison of common light types:

	LED	CFL	Incandescent
Lightbulb Projected Lifespan	50,000 hours	10,000 hours	1,200 hours
Watts Per Bulb (equiv. 60 watts)	10	14	60
Cost Per Bulb	Approx. $19.00	$7.00	$1.25
KWh of Electricity Used Over 50,000 Hours	500	700	3000
Cost of Electricity (@ 0.10 per KWh)	$50	$70	$300
Bulbs Needed for 50,000 Hours of Use	1	5	42
Equivalent 50,000 Hours Bulb Expense	$19.00	$35.00	$52.50
TOTAL Cost for 50,000 Hours	$69.00	$105.00	$352.50

Energy Consumption

In the US, one kilowatt-hour (3.6 MJ) of electricity currently causes an average 1.34 pounds (610 g) of CO_2 emission. Assuming the average light bulb is on for 10 hours a day, a 40-watt bulb will cause 196 pounds (89 kg) of CO_2 emission per year. The 6-watt LED equivalent will only cause 30 pounds (14 kg) of CO_2 over the same time span. A building's carbon footprint from lighting can therefore be reduced by 85% by exchanging all incandescent bulbs for new LEDs if a building previously used only incandescent bulbs.

In practice, most buildings that use a lot of lighting use fluorescent lighting, which has 22% luminous efficiency compared with 5% for filaments, so changing to LED lighting would still give a 34% reduction in electrical power use and carbon emissions.

The reduction in carbon emissions depends on the source of electricity. Nuclear power in the United States produced 19.2% of electricity in 2011, so reducing electricity consumption in the U.S. reduces carbon emissions more than in France (75% nuclear electricity) or Norway (almost entirely hydroelectric).

Replacing lights that spend the most time lit results in the most savings, so LED lights in infrequently used locations bring a smaller return on investment.

Light Sources for Machine Vision Systems

Machine vision systems often require bright and homogeneous illumination, so features of interest are easier to process. LEDs are often used for this purpose, and this is likely to remain one of their major uses until the price drops low enough to make signaling and illumination uses more widespread. Barcode scanners are the most common example of machine vision, and many low cost products use red LEDs instead of lasers. Optical computer mice are an example of LEDs in machine vision, as it is used to provide an even light source on the surface for the miniature camera within the mouse. LEDs constitute a nearly ideal light source for machine vision systems for several reasons:

- The size of the illuminated field is usually comparatively small and machine vision systems are often quite expensive, so the cost of the light source is usually a minor concern. However, it might not be easy to replace a broken light source placed within complex machinery, and here the long service life of LEDs is a benefit.

- LED elements tend to be small and can be placed with high density over flat or even-shaped substrates (PCBs etc.) so that bright and homogeneous sources that direct light from tightly controlled directions on inspected parts can be designed. This can often be obtained with small, low-cost lenses and diffusers, helping to achieve high light densities with control over lighting levels and homogeneity. LED sources can be shaped in several configurations (spot lights for reflective illumination; ring lights for coaxial illumination; back lights for contour illumination; linear assemblies; flat, large format panels; dome sources for diffused, omnidirectional illumination).

- LEDs can be easily strobed (in the microsecond range and below) and synchronized with imaging. High-power LEDs are available allowing well-lit images even with very short light pulses. This is often used to obtain crisp and sharp "still" images of quickly moving parts.

- LEDs come in several different colors and wavelengths, allowing easy use of the best color for each need, where different color may provide better visibility of features of interest. Having a precisely known spectrum allows tightly matched filters to be used to separate informative bandwidth or to reduce disturbing effects of ambient light. LEDs usually operate at comparatively low working temperatures, simplifying heat management and dissipation. This allows using plastic lenses, filters, and diffusers. Waterproof units can also easily be designed, allowing use in harsh or wet environments (food, beverage, oil industries).

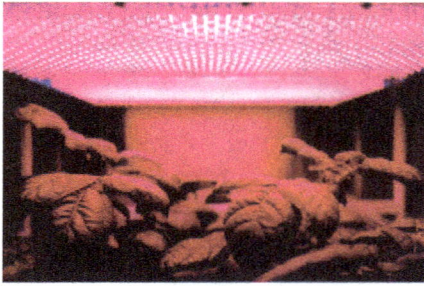

LED panel light source used in an experiment on plant growth. The findings of such experiments may be used to grow food in space on long duration missions.

Traffic light using LED

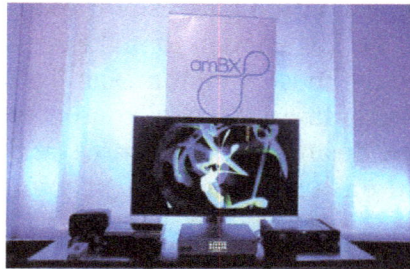

LED lights reacting dynamically to video feed via AmBX

Other Applications

The light from LEDs can be modulated very quickly so they are used extensively in optical fiber and free space optics communications. This includes remote controls, such as for TVs, VCRs, and LED Computers, where infrared LEDs are often used. Opto-isolators use an LED combined with a photodiode or phototransistor to provide a signal path with electrical isolation between two circuits. This is especially useful in medical equipment where the signals from a low-voltage sensor circuit (usually battery-powered) in contact with a living organism must be electrically isolated from any possible electrical failure in a recording or monitoring device operating at potentially dangerous voltages. An optoisolator also allows information to be transferred between circuits not sharing a common ground potential.

Many sensor systems rely on light as the signal source. LEDs are often ideal as a light source due to the requirements of the sensors. LEDs are used as motion sensors, for example in optical computer mice. The Nintendo Wii's sensor bar uses infrared LEDs. Pulse oximeters use them for measuring oxygen saturation. Some flatbed scanners use arrays of RGB LEDs rather than the typical cold-cathode fluorescent lamp as the light source. Having independent control of three illuminated colors allows the scanner to calibrate itself for more accurate color balance, and there is no need for warm-up. Further, its sensors only need be monochromatic, since at any one time the page being scanned is only lit by one color of light. Since LEDs can also be used as photodiodes, they can be used for both photo emission and detection. This could be used, for example, in a touchscreen that registers reflected light from a finger or stylus. Many materials and biological systems are sensitive to, or dependent on, light. Grow lights use LEDs to increase photosynthesis in plants, and bacteria and viruses can be removed from water and other substances using UV LEDs for sterilization.

LEDs have also been used as a medium-quality voltage reference in electronic circuits. The forward voltage drop (e.g. about 1.7 V for a normal red LED) can be used instead of a Zener diode in low-voltage regulators. Red LEDs have the flattest I/V curve above the knee. Nitride-based LEDs have a fairly steep I/V curve and are useless for this purpose. Although LED forward voltage is far more current-dependent than a Zener diode, Zener diodes with breakdown voltages below 3 V are not widely available.

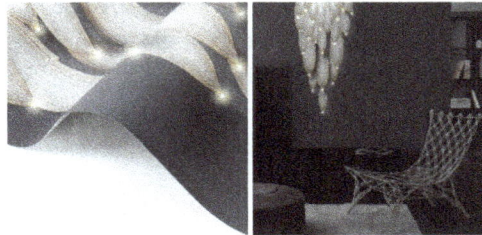

LED wallpaper by Meystyle

The progressive miniaturization of low-voltage lighting technology, such as LEDs and OLEDs, suitable to be incorporated into low-thickness materials has fostered in recent years the experimentation on combining light sources and wall covering surfaces to be applied onto interior walls. The new possibilities offered by these developments have prompted some designers and companies, such as Meystyle, Ingo Maurer, Lomox and Philips, to research and develop proprietary LED wallpaper technologies, some of which are currently available for commercial purchase. Other solutions mainly exist as prototypes or are in the process of being further refined.

Photodiode

A photodiode is a semiconductor device that converts light into current. The current is generated when photons are absorbed in the photodiode. A small amount of current is also produced when no light is present. Photodiodes may contain optical filters, built-in lenses, and may have large or small surface areas. Photodiodes usually have a slower response time as their surface area increases. The common, traditional solar cell used to generate electric solar power is a large area photodiode.

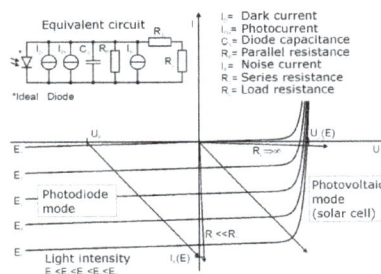

I-V characteristic of a photodiode. The linear load lines represent the response of the external circuit: I=(Applied bias voltage-Diode voltage)/Total resistance. The points of intersection with the curves represent the actual current and voltage for a given bias, resistance and illumination.

Photodiodes are similar to regular semiconductor diodes except that they may be either exposed (to detect vacuum UV or X-rays) or packaged with a window or optical fiber connection to allow light to reach the sensitive part of the device. Many diodes designed for use specifically as a photodiode use a PIN junction rather than a p–n junction, to increase the speed of response. A photodiode is designed to operate in reverse bias.

Principle of Operation

A photodiode is a p–n junction or PIN structure. When a photon of sufficient energy strikes the diode, it creates an electron-hole pair. This mechanism is also known as the inner photoelectric effect. If the absorption occurs in the junction's depletion region, or one diffusion length away from it, these carriers are swept from the junction by the built-in electric field of the depletion region. Thus holes move toward the anode, and electrons toward the cathode, and a photocurrent is produced. The total current through the photodiode is the sum of the dark current (current that is generated in the absence of light) and the photocurrent, so the dark current must be minimized to maximize the sensitivity of the device.

Photovoltaic Mode

When used in zero bias or *photovoltaic mode*, the flow of photocurrent out of the device is restricted and a voltage builds up. This mode exploits the photovoltaic effect, which is the basis for solar cells – a traditional solar cell is just a large area photodiode.

Photoconductive Mode

In this mode the diode is often reverse biased (with the cathode driven positive with respect to the anode). This reduces the response time because the additional reverse bias increases the width of the depletion layer, which decreases the junction's capacitance. The reverse bias also increases the dark current without much change in the photocurrent. For a given spectral distribution, the photocurrent is linearly proportional to the illuminance (and to the irradiance).

Although this mode is faster, the photoconductive mode tends to exhibit more electronic noise. The leakage current of a good PIN diode is so low (<1 nA) that the Johnson–Nyquist noise of the load resistance in a typical circuit often dominates.

Other Modes of Operation

Avalanche photodiodes are photodiodes with structure optimized for operating with high reverse bias, approaching the reverse breakdown voltage. This allows each *photo-generated* carrier to be multiplied by avalanche breakdown, resulting in internal gain within the photodiode, which increases the effective *responsivity* of the device.

Electronic symbol for a phototransistor

A phototransistor is a light-sensitive transistor. A common type of phototransistor, called a photobipolar transistor, is in essence a bipolar transistor encased in a transparent case so that light can reach the *base–collector junction*. It was invented by Dr. John N. Shive (more famous for his wave machine) at Bell Labs in 1948, but it was not announced until 1950. The electrons that are generated by photons in the base–collector junction are injected into the base, and this photodiode current is amplified by the transistor's current gain β (or h_{fe}). If the base and collector leads are used and the emitter is left unconnected, the phototransistor becomes a photodiode. While phototransistors have a higher responsivity for light they are not able to detect low levels of light any better than photodiodes.Phototransistors also have significantly longer response times. Field-effect phototransistors, also known as photoFETs, are light-sensitive field-effect transistors. Unlike photobipolar transistors, photoFETs control drain-source current by creating a gate voltage.

Materials

The material used to make a photodiode is critical to defining its properties, because only photons with sufficient energy to excite electrons across the material's bandgap will produce significant photocurrents.

Materials commonly used to produce photodiodes include:

Because of their greater bandgap, silicon-based photodiodes generate less noise than germanium-based photodiodes.

Unwanted Photodiode Effects

Any p–n junction, if illuminated, is potentially a photodiode. Semiconductor devices such as transistors and ICs contain p–n junctions, and will not function correctly if they are illuminated by unwanted electromagnetic radiation (light) of wavelength suitable to produce a photocurrent; this is avoided by encapsulating devices in opaque housings. If these housings are not completely opaque to high-energy radiation (ultraviolet, X-rays, gamma rays), transistors and ICs can malfunction due to induced photo-currents. Background radiation from the packaging is also significant. Radiation hardening mitigates these effects.

The infamous Raspberry Pi 2 xenon flash hanging bug is a result of xenon flash's intense ultraviolet emission lines disrupting a switch mode power supply controller chip in bare-die wafer-scale package, with the resulting power surge causing the main processor to lock up.

Features

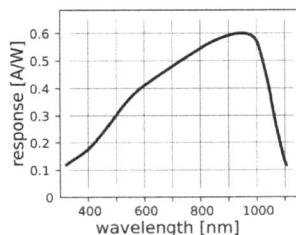

Response of a silicon photo diode vs wavelength of the incident light

Critical performance parameters of a photodiode include:

Responsivity

> The Spectral responsivity is a ratio of the generated photocurrent to incident light power, expressed in A/W when used in photoconductive mode. The wavelength-dependence may also be expressed as a *Quantum efficiency*, or the ratio of the number of photogenerated carriers to incident photons, a unitless quantity.

Dark Current

> The current through the photodiode in the absence of light, when it is operated in photoconductive mode. The dark current includes photocurrent generated by background radiation and the saturation current of the semiconductor junction. Dark current must be accounted for by calibration if a photodiode is used to make an accurate optical power measurement, and it is also a source of noise when a photodiode is used in an optical communication system.

Response Time

> A photon absorbed by the semiconducting material will generate an electron-hole pair which will in turn start moving in the material under the effect of the electric field and thus generate a current. The finite duration of this current is known as the transit-time spread and can be evaluated by using Ramo's theorem. One can also show with this theorem that the total charge generated in the external circuit is well e and not 2e as might seem by the presence of the two carriers. Indeed, the integral of the current due to both electron and hole over time must be equal to e. The resistance and capacitance of the photodiode and the external circuitry give rise to another response time known as RC time constant $\tau = RC$. This combination of R and C integrates the photoresponse over time and thus lengthens the impulse response of the photodiode. When used in an optical communication system, the response time determines the bandwidth available for signal modulation and thus data transmission.

Noise-equivalent Power

> (NEP) The minimum input optical power to generate photocurrent, equal to the rms noise current in a 1 hertz bandwidth. NEP is essentially the minimum detectable power. The related characteristic detectivity (D) is the inverse of NEP, 1/NEP. There is also the specific detectivity ($D^\star$) which is the detectivity multiplied by the square root of the area (A) of the photodetector, ($D^\star = D\sqrt{A}$) for a 1 Hz bandwidth. The specific detectivity allows different systems to be compared independent of sensor area and system bandwidth; a higher detectivity value indicates a low-noise device or system. Although it is traditional to give ($D^\star$) in many catalogues as a measure of the diode's quality, in practice, it is hardly ever the key parameter.

When a photodiode is used in an optical communication system, all these parameters contribute to the *sensitivity* of the optical receiver, which is the minimum input power required for the receiver to achieve a specified *bit error rate*.

Applications

P–n photodiodes are used in similar applications to other photodetectors, such as photoconductors, charge-coupled devices, and photomultiplier tubes. They may be used to generate an output which is dependent upon the illumination (analog; for measurement and the like), or to change the state of circuitry (digital; either for control and switching, or digital signal processing).

Photodiodes are used in consumer electronics devices such as compact disc players, smoke detectors, and the receivers for infrared remote control devices used to control equipment from televisions to air conditioners. For many applications either photodiodes or photoconductors may be used. Either type of photosensor may be used for light measurement, as in camera light meters, or to respond to light levels, as in switching on street lighting after dark.

Photosensors of all types may be used to respond to incident light, or to a source of light which is part of the same circuit or system. A photodiode is often combined into a single component with an emitter of light, usually a light-emitting diode (LED), either to detect the presence of a mechanical obstruction to the beam (slotted optical switch), or to couple two digital or analog circuits while maintaining extremely high electrical isolation between them, often for safety (optocoupler). The combination of LED and photodiode is also used in many sensor systems to characterize different types of products based on their optical absorbance.

Photodiodes are often used for accurate measurement of light intensity in science and industry. They generally have a more linear response than photoconductors.

They are also widely used in various medical applications, such as detectors for computed tomography (coupled with scintillators), instruments to analyze samples (immunoassay), and pulse oximeters.

PIN diodes are much faster and more sensitive than p–n junction diodes, and hence are often used for optical communications and in lighting regulation.

P–n photodiodes are not used to measure extremely low light intensities. Instead, if high sensitivity is needed, avalanche photodiodes, intensified charge-coupled devices or photomultiplier tubes are used for applications such as astronomy, spectroscopy, night vision equipment and laser rangefinding.

Pinned photodiode is not a PIN photodiode, it has p+/n/p regions in it. It has a shallow P+ implant in N type diffusion layer over a P-type epitaxial substrate layer. It is used in CMOS Active pixel sensor.

Comparison With Photomultipliers

Advantages compared to photomultipliers:

1. Excellent linearity of output current as a function of incident light

2. Spectral response from 190 nm to 1100 nm (silicon), longer wavelengths with other semiconductor materials

3. Low noise

4. Ruggedized to mechanical stress

5. Low cost

6. Compact and light weight

7. Long lifetime

8. High quantum efficiency, typically 60–80%

9. No high voltage required

Disadvantages Compared To Photomultipliers:

1. Small area

2. No internal gain (except avalanche photodiodes, but their gain is typically 10^2–10^3 compared to 10^5-10^8 for the photomultiplier)

3. Much lower overall sensitivity

4. Photon counting only possible with specially designed, usually cooled photodiodes, with special electronic circuits

5. Response time for many designs is slower

6. latent effect

Photodiode Array

A one-dimensional array of hundreds or thousands of photodiodes can be used as a position sensor, for example as part of an angle sensor. One advantage of photodiode arrays (PDAs) is that they allow for high speed parallel read out since the driving electronics may not be built in like a traditional CMOS or CCD sensor.

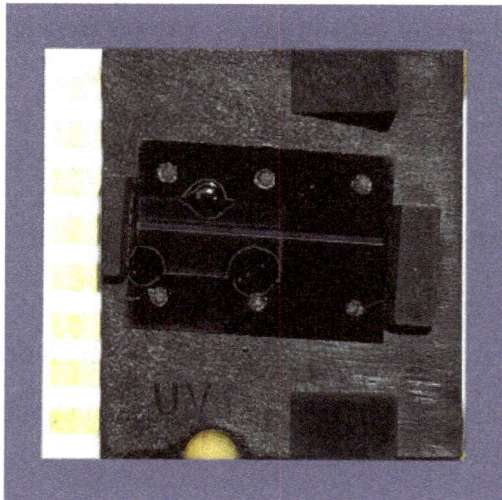

A 2 x 2 cm photodiode array chip with more than 200 diodes

Solar Cell

A solar cell, or photovoltaic cell in very early days also termed "solar battery" – a denotation which nowadays has a totally different meaning, is an electrical device that converts the energy of light directly into electricity by the photovoltaic effect, which is a physical and chemical phenomenon. It is a form of photoelectric cell, defined as a device whose electrical characteristics, such as current, voltage, or resistance, vary when exposed to light. Solar cells are the building blocks of photovoltaic modules, otherwise known as solar panels.

Solar cells are described as being photovoltaic irrespective of whether the source is sunlight or an artificial light. They are used as a photodetector (for example infrared detectors), detecting light or other electromagnetic radiation near the visible range, or measuring light intensity.

A conventional crystalline silicon solar cell (as of 2005). Electrical contacts made from busbars (the larger silver-colored strips) and fingers (the smaller ones) are printed on the silicon wafer.

The operation of a photovoltaic (PV) cell requires 3 basic attributes:

- The absorption of light, generating either electron-hole pairs or excitons.
- The separation of charge carriers of opposite types.
- The separate extraction of those carriers to an external circuit.

In contrast, a solar thermal collector supplies heat by absorbing sunlight, for the purpose of either direct heating or indirect electrical power generation from heat. A "photoelectrolytic cell" (photo-electrochemical cell), on the other hand, refers either to a type of photovoltaic cell (like that developed by Edmond Becquerel and modern dye-sensitized solar cells), or to a device that splits water directly into hydrogen and oxygen using only solar illumination.

Applications

Assemblies of solar cells are used to make solar modules which generate electrical power from sunlight, as distinguished from a "solar thermal module" or "solar hot water panel". A solar array generates solar power using solar energy.

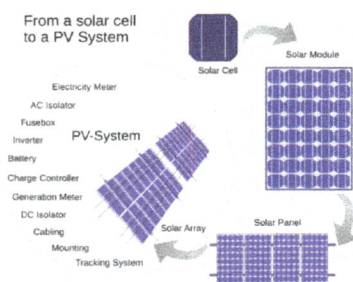

From a solar cell to a PV system. Diagram of the possible components of a photovoltaic system

Cells, Modules, Panels and Systems

Multiple solar cells in an integrated group, all oriented in one plane, constitute a solar photovoltaic panel or solar photovoltaic module. Photovoltaic modules often have a sheet of glass on the sun-facing side, allowing light to pass while protecting the semiconductor wafers. Solar cells are usually connected in series and parallel circuits or series in modules, creating an additive voltage. Connecting cells in parallel yields a higher current; however, problems such as shadow effects can shut down the weaker (less illuminated) parallel string (a number of series connected cells) causing substantial power loss and possible damage because of the reverse bias applied to the shadowed cells by their illuminated partners. Strings of series cells are usually handled independently and not connected in parallel, though as of 2014, individual power boxes are often supplied for each module, and are connected in parallel. Although modules can be interconnected to create an array with the desired peak DC voltage and loading current capacity, using independent MPPTs (maximum power point trackers) is preferable. Otherwise, shunt diodes can reduce shadowing power loss in arrays with series/parallel connected cells.

Typical PV system prices in 2013 in selected countries (USD)								
USD/W	Australia	China	France	Germany	Italy	Japan	United Kingdom	United States
Residential	1.8	1.5	4.1	2.4	2.8	4.2	2.8	4.9
Commercial	1.7	1.4	2.7	1.8	1.9	3.6	2.4	4.5
Utility-scale	2.0	1.4	2.2	1.4	1.5	2.9	1.9	3.3
Source: *IEA – Technology Roadmap: Solar Photovoltaic Energy report*, 2014 edition Note: *DOE – Photovoltaic System Pricing Trends* reports lower prices for the U.S.								

History

The photovoltaic effect was experimentally demonstrated first by French physicist Edmond Becquerel. In 1839, at age 19, he built the world's first photovoltaic cell in his father's laboratory. Willoughby Smith first described the "Effect of Light on Selenium during the passage of an Electric Current" in a 20 February 1873 issue of Nature. In 1883 Charles Fritts built the first solid state photovoltaic cell by coating the semiconductor selenium with a thin layer of gold to form the junctions; the device was only around 1% efficient.

In 1888 Russian physicist Aleksandr Stoletov built the first cell based on the outer photoelectric effect discovered by Heinrich Hertz in 1887.

In 1905 Albert Einstein proposed a new quantum theory of light and explained the photoelectric effect in a landmark paper, for which he received the Nobel Prize in Physics in 1921.

Vadim Lashkaryov discovered *p-n*-junctions in CuO and silver sulphide protocells in 1941.

Russell Ohl patented the modern junction semiconductor solar cell in 1946 while working on the series of advances that would lead to the transistor.

The first practical photovoltaic cell was publicly demonstrated on 25 April 1954 at Bell Laboratories. The inventors were Daryl Chapin, Calvin Souther Fuller and Gerald Pearson.

Solar cells gained prominence with their incorporation onto the 1958 Vanguard I satellite.

Improvements were gradual over the next two decades. However, this success was also the reason that costs remained high, because space users were willing to pay for the best possible cells, leaving no reason to invest in lower-cost, less-efficient solutions. The price was determined largely by the semiconductor industry; their move to integrated circuits in the 1960s led to the availability of larger boules at lower relative prices. As their price fell, the price of the resulting cells did as well. These effects lowered 1971 cell costs to some $100 per watt.

Space Applications

Solar cells were first used in a prominent application when they were proposed and flown on the Vanguard satellite in 1958, as an alternative power source to the primary battery power source. By adding cells to the outside of the body, the mission time could be extended with no major changes to the spacecraft or its power systems. In 1959 the United States launched Explorer 6, featuring large wing-shaped solar arrays, which became a common feature in satellites. These arrays consisted of 9600 Hoffman solar cells.

By the 1960s, solar cells were (and still are) the main power source for most Earth orbiting satellites and a number of probes into the solar system, since they offered the best power-to-weight ratio. However, this success was possible because in the space application, power system costs could be high, because space users had few other power options, and were willing to pay for the best possible cells. The space power market drove the development of higher efficiencies in solar cells up until the National Science Foundation "Research Applied to National Needs" program began to push development of solar cells for terrestrial applications.

In the early 1990s the technology used for space solar cells diverged from the silicon technology used for terrestrial panels, with the spacecraft application shifting to gallium arsenide-based III-V semiconductor materials, which then evolved into the modern III-V multijunction photovoltaic cell used on spacecraft.

Price Reductions

In late 1969 Elliot Berman joined Exxon's task force which was looking for projects 30 years in the future and in April 1973 he founded Solar Power Corporation, a wholly owned subsidiary of Exxon at that time. The group had concluded that electrical power would be much more expensive by 2000, and felt that this increase in price would make alternative energy sources more attractive. He conducted a market study and concluded that a price per watt of about $20/watt would create significant demand. The team eliminated the steps of polishing the wafers and coating them with an anti-reflective layer, relying on the rough-sawn wafer surface. The team also replaced the expensive materials and hand wiring used in space applications with a printed circuit board on the back, acrylic plastic on the front, and silicone glue between the two, "potting" the cells. Solar cells could be made using cast-off material from the electronics market. By 1973 they announced a product, and SPC convinced Tideland Signal to use its panels to power navigational buoys, initially for the U.S. Coast Guard.

Research into solar power for terrestrial applications became prominent with the U.S. National Science Foundation's Advanced Solar Energy Research and Development Division within the "Research Applied to National Needs" program, which ran from 1969 to 1977, and funded research on developing solar power for ground electrical power systems. A 1973 conference, the "Cherry Hill Conference", set forth the technology goals required to achieve this goal and outlined an ambitious project for achieving them, kicking off an applied research program that would be ongoing for several decades. The program was eventually taken over by the Energy Research and Development Administration (ERDA), which was later merged into the U.S. Department of Energy.

Following the 1973 oil crisis, oil companies used their higher profits to start (or buy) solar firms, and were for decades the largest producers. Exxon, ARCO, Shell, Amoco (later purchased by BP) and Mobil all had major solar divisions during the 1970s and 1980s. Technology companies also participated, including General Electric, Motorola, IBM, Tyco and RCA.

Declining Costs and Exponential Growth

Adjusting for inflation, it cost $96 per watt for a solar module in the mid-1970s. Process improvements and a very large boost in production have brought that figure down 99 percent, to 68¢ per watt in 2016, according to data from Bloomberg New Energy Finance. Swanson's law is an observation similar to Moore's Law that states that solar cell prices fall 20% for every doubling of industry capacity. It was featured in an article in the British weekly newspaper The Economist.

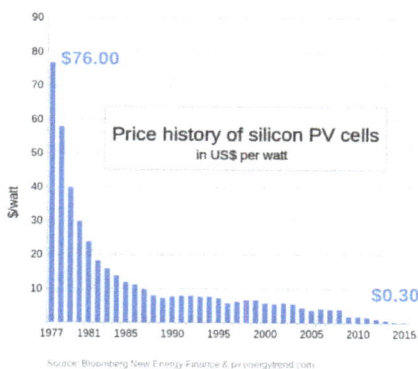

Price per watt history for conventional (c-Si) solar cells since 1977

Further improvements reduced production cost to under $1 per watt, with wholesale costs well under $2. Balance of system costs were then higher than the panels. Large commercial arrays could be built, as of 2010, at below $3.40 a watt, fully commissioned.

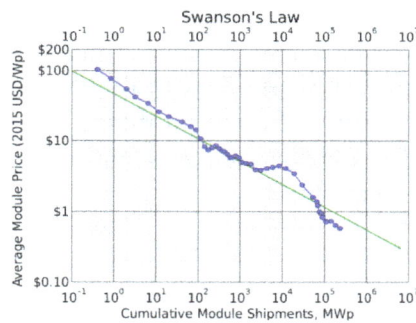

Swanson's law – the learning curve of solar PV

As the semiconductor industry moved to ever-larger boules, older equipment became inexpensive. Cell sizes grew as equipment became available on the surplus market; ARCO Solar's original panels used cells 2 to 4 inches (50 to 100 mm) in diameter. Panels in the 1990s and early 2000s generally used 125 mm wafers; since 2008 almost all new panels use 150 mm cells. The widespread introduction of flat screen televisions in the late 1990s and early 2000s led to the wide availability of large, high-quality glass sheets to cover the panels.

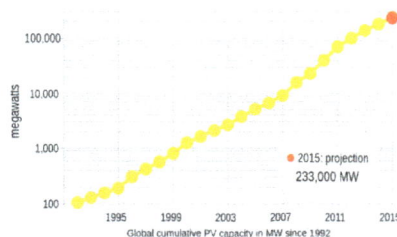

Growth of photovoltaics – Worldwide total installed PV capacity

During the 1990s, polysilicon ("poly") cells became increasingly popular. These cells offer less efficiency than their monosilicon ("mono") counterparts, but they are grown in large vats that reduce cost. By the mid-2000s, poly was dominant in the low-cost panel market, but more recently the mono returned to widespread use.

Manufacturers of wafer-based cells responded to high silicon prices in 2004–2008 with rapid reductions in silicon consumption. In 2008, according to Jef Poortmans, director of IMEC's organic and solar department, current cells use 8–9 grams (0.28–0.32 oz) of silicon per watt of power generation, with wafer thicknesses in the neighborhood of 200 microns. Crystalline silicon panels dominate worldwide markets and are mostly manufactured in China and Taiwan. By late 2011, a drop in European demand due to budgetary turmoil dropped prices for crystalline solar modules to about $1.09 per watt down sharply from 2010. Prices continued to fall in 2012, reaching $0.62/watt by 4Q2012.

Global installed PV capacity reached at least 177 gigawatts in 2014, enough to supply 1 percent of the world's total electricity consumption. Solar PV is growing fastest in Asia, with China and Japan currently accounting for half of worldwide deployment.

Subsidies and grid parity

Solar-specific feed-in tariffs vary by country and within countries. Such tariffs encourage the development of solar power projects. Widespread grid parity, the point at which photovoltaic electricity is equal to or cheaper than grid power without subsidies, likely requires advances on all three fronts. Proponents of solar hope to achieve grid parity first in areas with abundant sun and high electricity costs such as in California and Japan. In 2007 BP claimed grid parity for Hawaii and other islands that otherwise use diesel fuel to produce electricity. George W. Bush set 2015 as the date for grid parity in the US. The Photovoltaic Association reported in 2012 that Australia had reached grid parity (ignoring feed in tariffs).

The price of solar panels fell steadily for 40 years, interrupted in 2004 when high subsidies in Germany drastically increased demand there and greatly increased the price of purified silicon (which is used in computer chips as well as solar panels). The recession of 2008 and the onset of Chinese manufacturing caused prices to resume their decline. In the four years after January 2008 prices for solar modules in Germany dropped from €3 to €1 per peak watt. During that same time production capacity surged with an annual growth of more than 50%. China increased market share from 8% in 2008 to over 55% in the last quarter of 2010. In December 2012 the price of Chinese solar panels had dropped to $0.60/Wp (crystalline modules).

Theory

The solar cell works in several steps:

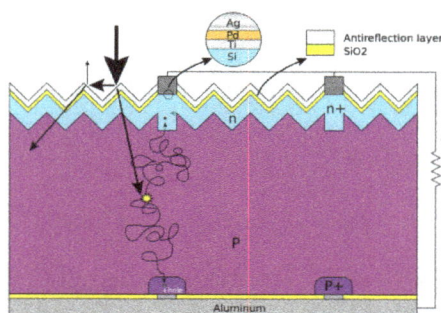

Working mechanism of a solar cell

- Photons in sunlight hit the solar panel and are absorbed by semiconducting materials, such as silicon.

- Electrons are excited from their current molecular/atomic orbital. Once excited an electron can either dissipate the energy as heat and return to its orbital or travel through the cell until it reaches an electrode. Current flows through the material to cancel the potential and this electricity is captured. The chemical bonds of the material are vital for this process to work, and usually silicon is used in two layers, one layer being bonded with boron, the other phosphorus. These layers have different chemical electric charges and subsequently both drive and direct the current of electrons.

- An array of solar cells converts solar energy into a usable amount of direct current (DC) electricity.

- An inverter can convert the power to alternating current (AC).

The most commonly known solar cell is configured as a large-area p–n junction made from silicon.

Efficiency

Solar cell efficiency may be broken down into reflectance efficiency, thermodynamic efficiency, charge carrier separation efficiency and conductive efficiency. The overall efficiency is the product of these individual metrics.

A solar cell has a voltage dependent efficiency curve, temperature coefficients, and allowable shadow angles.

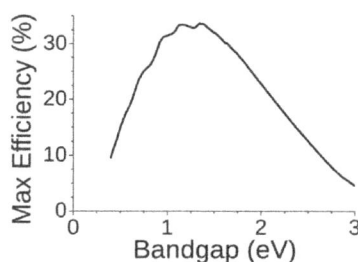

The Shockley-Queisser limit for the theoretical maximum efficiency of a solar cell. Semiconductors with band gap between 1 and 1.5eV, or near-infrared light, have the greatest potential to form an efficient single-junction cell. (The efficiency "limit" shown here can be exceeded by multijunction solar cells.)

Due to the difficulty in measuring these parameters directly, other parameters are substituted: thermodynamic efficiency, quantum efficiency, integrated quantum efficiency, V_{OC} ratio, and fill factor. Reflectance losses are a portion of quantum efficiency under "external quantum efficiency". Recombination losses make up another portion of quantum efficiency, V_{OC} ratio, and fill factor. Resistive losses are predominantly categorized under fill factor, but also make up minor portions of quantum efficiency, V_{OC} ratio.

The fill factor is the ratio of the actual maximum obtainable power to the product of the open circuit voltage and short circuit current. This is a key parameter in evaluating performance. In 2009, typical commercial solar cells had a fill factor > 0.70. Grade B cells were usually between 0.4 and 0.7. Cells with a high fill factor have a low equivalent series resistance and a high equivalent shunt resistance, so less of the current produced by the cell is dissipated in internal losses.

Single p–n junction crystalline silicon devices are now approaching the theoretical limiting power efficiency of 33.7%, noted as the Shockley–Queisser limit in 1961. In the extreme, with an infinite number of layers, the corresponding limit is 86% using concentrated sunlight.

In December 2014, a solar cell achieved a new laboratory record with 46 percent efficiency in a French-German collaboration.

In 2014, three companies broke the record of 25.6% for a silicon solar cell. Panasonic's was the most efficient. The company moved the front contacts to the rear of the panel, eliminating shaded areas. In addition they applied thin silicon films to the (high quality silicon) wafer's front and back to eliminate defects at or near the wafer surface.

In September 2015, the Fraunhofer Institute for Solar Energy Systems (Fraunhofer ISE) announced the achievement of an efficiency above 20% for epitaxial wafer cells. The work on optimizing the atmospheric-pressure chemical vapor deposition (APCVD) in-line production chain was done in collaboration with NexWafe GmbH, a company spun off from Fraunhofer ISE to commercialize production.

For triple-junction thin-film solar cells, the world record is 13.6%, set in June 2015.

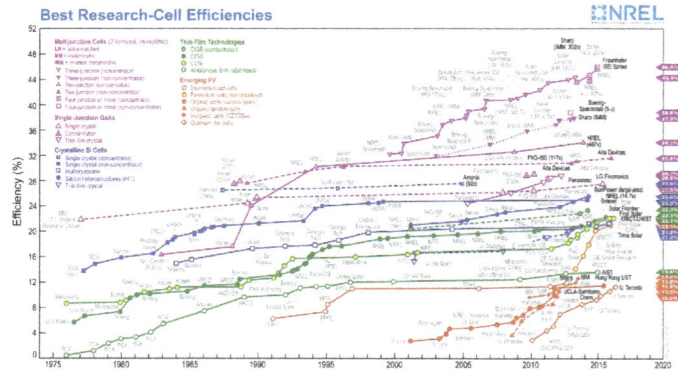

Reported timeline of solar cell energy conversion efficiencies (National Renewable Energy Laboratory)

Materials

Solar cells are typically named after the semiconducting material they are made of. These materials must have certain characteristics in order to absorb sunlight. Some cells are designed to handle sunlight that reaches the Earth's surface, while others are optimized for use in space. Solar cells can be made of only one single layer of light-absorbing material (single-junction) or use multiple physical configurations (multi-junctions) to take advantage of various absorption and charge separation mechanisms.

Solar cells can be classified into first, second and third generation cells. The first generation cells—also called conventional, traditional or wafer-based cells—are made of crystalline silicon, the commercially predominant PV technology, that includes materials such as polysilicon and monocrystalline silicon. Second generation cells are thin film solar cells, that include amorphous silicon, CdTe and CIGS cells and are commercially significant in utility-scale photovoltaic power stations, building integrated photovoltaics or in small stand-alone power system. The third generation of solar cells includes a number of thin-film technologies often described as emerging photovoltaics—most of them have not yet been commercially applied and are still in the research or development phase. Many use organic materials, often organometallic compounds as well as inorganic substances. Despite the fact that their efficiencies had been low and the stability of the absorber material was often too short for commercial applications, there is a lot of research invested into these technologies as they promise to achieve the goal of producing low-cost, high-efficiency solar cells.

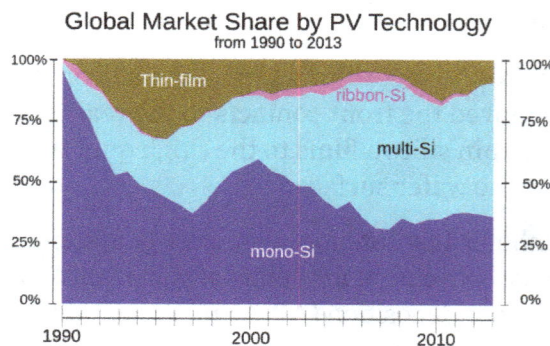

Global market-share in terms of annual production by PV technology since 1990

Crystalline Silicon

By far, the most prevalent bulk material for solar cells is crystalline silicon (c-Si), also known as "solar grade silicon". Bulk silicon is separated into multiple categories according to crystallinity and crystal size in the resulting ingot, ribbon or wafer. These cells are entirely based around the concept of a p-n junction. Solar cells made of c-Si are made from wafers between 160 and 240 micrometers thick.

Monocrystalline Silicon

Monocrystalline silicon (mono-Si) solar cells are more efficient and more expensive than most other types of cells. The corners of the cells look clipped, like an octagon, because the wafer material is cut from cylindrical ingots, that are typically grown by the Czochralski process. Solar panels using mono-Si cells display a distinctive pattern of small white diamonds.

Epitaxial Silicon

Epitaxial wafers can be grown on a monocrystalline silicon "seed" wafer by atmospheric-pressure CVD in a high-throughput inline process, and then detached as self-supporting wafers of some standard thickness (e.g., 250 μm) that can be manipulated by hand, and directly substituted for wafer cells cut from monocrystalline silicon ingots. Solar cells made with this technique can have efficiencies approaching those of wafer-cut cells, but at appreciably lower cost.

Polycrystalline Silicon

Polycrystalline silicon, or multicrystalline silicon (multi-Si) cells are made from cast square ingots—large blocks of molten silicon carefully cooled and solidified. They consist of small crystals giving the material its typical metal flake effect. Polysilicon cells are the most common type used in photovoltaics and are less expensive, but also less efficient, than those made from monocrystalline silicon.

Ribbon Silicon

Ribbon silicon is a type of polycrystalline silicon—it is formed by drawing flat thin films from molten silicon and results in a polycrystalline structure. These cells are cheaper to make than multi-Si, due to a great reduction in silicon waste, as this approach does not require sawing from ingots. However, they are also less efficient.

Mono-like-multi Silicon (MLM)

This form was developed in the 2000s and introduced commercially around 2009. Also called cast-mono, this design uses polycrystalline casting chambers with small "seeds" of mono material. The result is a bulk mono-like material that is polycrystalline around the outsides. When sliced for processing, the inner sections are high-efficiency mono-like cells (but square instead of "clipped"), while the outer edges are sold as conventional poly. This production method results in mono-like cells at poly-like prices.

Thin Film

Thin-film technologies reduce the amount of active material in a cell. Most designs sandwich active material between two panes of glass. Since silicon solar panels only use one pane of glass, thin film panels are approximately twice as heavy as crystalline silicon panels, although they have a smaller ecological impact (determined from life cycle analysis).

Cadmium Telluride

Cadmium telluride is the only thin film material so far to rival crystalline silicon in cost/watt. However cadmium is highly toxic and tellurium (anion: "telluride") supplies are limited. The cadmium present in the cells would be toxic if released. However, release is impossible during normal operation of the cells and is unlikely during fires in residential roofs. A square meter of CdTe contains approximately the same amount of Cd as a single C cell nickel-cadmium battery, in a more stable and less soluble form.

Copper Indium Gallium Selenide

Copper indium gallium selenide (CIGS) is a direct band gap material. It has the highest efficiency (~20%) among all commercially significant thin film materials (CIGS solar cell). Traditional methods of fabrication involve vacuum processes including co-evaporation and sputtering. Recent developments at IBM and Nanosolar attempt to lower the cost by using non-vacuum solution processes.

Silicon Thin Film

Silicon thin-film cells are mainly deposited by chemical vapor deposition (typically plasma-enhanced, PE-CVD) from silane gas and hydrogen gas. Depending on the deposition parameters, this can yield amorphous silicon (a-Si or a-Si:H), protocrystalline silicon or nanocrystalline silicon (nc-Si or nc-Si:H), also called microcrystalline silicon.

Amorphous silicon is the most well-developed thin film technology to-date. An amorphous silicon (a-Si) solar cell is made of non-crystalline or microcrystalline silicon. Amorphous silicon has a higher bandgap (1.7 eV) than crystalline silicon (c-Si) (1.1 eV), which means it absorbs the visible part of the solar spectrum more strongly than the higher power density infrared portion of the spectrum. The production of a-Si thin film solar cells uses glass as a substrate and deposits a very thin layer of silicon by plasma-enhanced chemical vapor deposition (PECVD).

Protocrystalline silicon with a low volume fraction of nanocrystalline silicon is optimal for high open circuit voltage. Nc-Si has about the same bandgap as c-Si and nc-Si and a-Si can advantageously be combined in thin layers, creating a layered cell called a tandem cell. The top cell in a-Si absorbs the visible light and leaves the infrared part of the spectrum for the bottom cell in nc-Si.

Gallium Arsenide Thin Film

The semiconductor material Gallium arsenide (GaAs) is also used for single-crystalline thin film solar cells. Although GaAs cells are very expensive, they hold the world's record in efficiency for a single-junction solar cell at 28.8%. GaAs is more commonly used in multi-junction photovoltaic cells for concentrated photovoltaics (CPV, HCPV) and for solar panels on spacecrafts, as the industry favours efficiency over cost for space-based solar power.

Multijunction Cells

Multi-junction cells consist of multiple thin films, each essentially a solar cell grown on top of another, typically using metalorganic vapour phase epitaxy. Each layers has a different band gap energy to allow it to absorb electromagnetic radiation over a different portion of the spectrum. Multi-junction cells were originally developed for special applications such as satellites and space exploration, but are now used increasingly in terrestrial concentrator photovoltaics (CPV), an emerging technology that uses lenses and curved mirrors to concentrate sunlight onto small but highly efficient multi-junction solar cells. By concentrating sunlight up to a thousand times, *High concentrated photovoltaics (HCPV)* has the potential to outcompete conventional solar PV in the future.

Dawn's 10 kW triple-junction gallium arsenide solar array at full extension

Tandem solar cells based on monolithic, series connected, gallium indium phosphide (GaInP), gallium arsenide (GaAs), and germanium (Ge) p–n junctions, are increasing sales, despite cost pressures. Between December 2006 and December 2007, the cost of 4N gallium metal rose from about \$350 per kg to \$680 per kg. Additionally, germanium metal prices have risen substantially to \$1000–1200 per kg this year. Those materials include gallium (4N, 6N and 7N Ga), arsenic (4N, 6N and 7N) and germanium, pyrolitic boron nitride (pBN) crucibles for growing crystals, and boron oxide, these products are critical to the entire substrate manufacturing industry.

A triple-junction cell, for example, may consist of the semiconductors: GaAs, Ge, and GaInP 2. Triple-junction GaAs solar cells were used as the power source of the Dutch four-time World Solar Challenge winners Nuna in 2003, 2005 and 2007 and by the Dutch solar cars Solutra (2005), Twente One (2007) and 21Revolution (2009).GaAs based multi-junction devices are the most efficient solar cells to date. On 15 October 2012, triple junction metamorphic cells reached a record high of 44%.

Research in Solar Cells

Perovskite Solar Cells

Perovskite solar cells are solar cells that include a perovskite-structured material as the active layer. Most commonly, this is a solution-processed hybrid organic-inorganic tin or lead halide based material. Efficiencies have increased from below 5% at their first usage in 2009 to over 20% in 2014, making them a very rapidly advancing technology and a hot topic in the solar cell field. Perovskite solar cells are also forecast to be extremely cheap to scale up, making them a very attractive option for commercialisation.

Liquid Inks

In 2014, researchers at California NanoSystems Institute discovered using kesterite and perovskite improved electric power conversion efficiency for solar cells.

Upconversion and Downconversion

Photon upconversion is the process of using two low-energy (*e.g.*, infrared) photons to produce one higher energy photon; downconversion is the process of using one high energy photon (*e.g.*,, ultraviolet) to produce two lower energy photons. Either of these techniques could be used to produce higher efficiency solar cells by allowing solar photons to be more efficiently used. The difficulty, however, is that the conversion efficiency of existing phosphors exhibiting up- or down-conversion is low, and is typically narrow band.

One upconversion technique is to incorporate lanthanide-doped materials(Er^{3+}, Yb^{3+}, Ho^{3+} or a combination), taking advantage of their luminescence to convert infrared radiation to visible light. Upconversion process occurs when two infrared photons are absorbed by rare-earth ions to generate a (high-energy) absorbable photon. As example, the energy transfer upconversion process (ETU), consists in successive transfer processes between excited ions in the near infrared. The upconverter material could be placed below the solar cell to absorb the infrared light that passes through the silicon. Useful ions are most commonly found in the trivalent state. Er^+ ions have been the most used. Er^{3+} ions absorb solar radiation around 1.54 μm. Two Er^{3+} ions that have absorbed this radiation can interact with each other through an upconversion process. The excited ion emits light above the Si bandgap that is absorbed by the solar cell and creates an additional electron–hole pair that can generate current. However, the increased efficiency was small. In addition, fluoroindate glasses have low phonon energy and have been proposed as suitable matrix doped with Ho^{3+} ions.

Light-absorbing Dyes

Dye-sensitized solar cells (DSSCs) are made of low-cost materials and do not need elaborate manufacturing equipment, so they can be made in a DIY fashion. In bulk it should be significantly less expensive than older solid-state cell designs. DSSC's can be engineered into flexible sheets and although its conversion efficiency is less than the best thin film cells, its price/performance ratio may be high enough to allow them to compete with fossil fuel electrical generation.

Typically a ruthenium metalorganic dye (Ru-centered) is used as a monolayer of light-absorbing material. The dye-sensitized solar cell depends on a mesoporous layer of nanoparticulate titanium dioxide to greatly amplify the surface area (200–300 m²/g TiO_2, as compared to approximately 10 m²/g of flat single crystal). The photogenerated electrons from the light absorbing dye are passed on to the n-type TiO_2 and the holes are absorbed by an electrolyte on the other side of the dye. The circuit is completed by a redox couple in the electrolyte, which can be liquid or solid. This type of cell allows more flexible use of materials and is typically manufactured by screen printing or ultrasonic nozzles, with the potential for lower processing costs than those used for bulk solar cells. However, the dyes in these cells also suffer from degradation under heat and UV light and the cell casing is difficult to seal due to the solvents used in assembly. The first commercial shipment of DSSC solar modules occurred in July 2009 from G24i Innovations.

Quantum Dots

Quantum dot solar cells (QDSCs) are based on the Gratzel cell, or dye-sensitized solar cell architecture, but employ low band gap semiconductor nanoparticles, fabricated with crystallite sizes small enough to form quantum dots (such as CdS, CdSe, Sb2S3, PbS, etc.), instead of organic or organometallic dyes as light absorbers. QD's size quantization allows for the band gap to be tuned by simply changing particle size. They also have high extinction coefficients and have shown the possibility of multiple exciton generation.

In a QDSC, a mesoporous layer of titanium dioxide nanoparticles forms the backbone of the cell, much like in a DSSC. This TiO2 layer can then be made photoactive by coating with semiconductor quantum dots using chemical bath deposition, electrophoretic deposition or successive ionic layer adsorption and reaction. The electrical circuit is then completed through the use of a liquid or solid redox couple. The efficiency of QDSCs has increased to over 5% shown for both liquid-junction and solid state cells. In an effort to decrease production costs, the Prashant Kamat research group demonstrated a solar paint made with TiO2 and CdSe that can be applied using a one-step method to any conductive surface with efficiencies over 1%.

Organic/Polymer Solar Cells

Organic solar cells and polymer solar cells are built from thin films (typically 100 nm) of organic semiconductors including polymers, such as polyphenylene vinylene and small-molecule compounds like copper phthalocyanine (a blue or green organic pigment) and carbon fullerenes and fullerene derivatives such as PCBM.

They can be processed from liquid solution, offering the possibility of a simple roll-to-roll printing process, potentially leading to inexpensive, large-scale production. In addition, these cells could be beneficial for some applications where mechanical flexibility and disposability are important. Current cell efficiencies are, however, very low, and practical devices are essentially non-existent.

Energy conversion efficiencies achieved to date using conductive polymers are very low compared to inorganic materials. However, Konarka Power Plastic reached efficiency of 8.3% and organic tandem cells in 2012 reached 11.1%.

The active region of an organic device consists of two materials, one electron donor and one electron acceptor. When a photon is converted into an electron hole pair, typically in the donor material, the charges tend to remain bound in the form of an exciton, separating when the exciton diffuses to the donor-acceptor interface, unlike most other solar cell types. The short exciton diffusion lengths of most polymer systems tend to limit the efficiency of such devices. Nanostructured interfaces, sometimes in the form of bulk heterojunctions, can improve performance.

In 2011, MIT and Michigan State researchers developed solar cells with a power efficiency close to

2% with a transparency to the human eye greater than 65%, achieved by selectively absorbing the ultraviolet and near-infrared parts of the spectrum with small-molecule compounds. Researchers at UCLA more recently developed an analogous polymer solar cell, following the same approach, that is 70% transparent and has a 4% power conversion efficiency. These lightweight, flexible cells can be produced in bulk at a low cost and could be used to create power generating windows.

In 2013, researchers announced polymer cells with some 3% efficiency. They used block copolymers, self-assembling organic materials that arrange themselves into distinct layers. The research focused on P3HT-b-PFTBT that separates into bands some 16 nanometers wide.

Adaptive Cells

Adaptive cells change their absorption/reflection characteristics depending to respond to environmental conditions. An adaptive material responds to the intensity and angle of incident light. At the part of the cell where the light is most intense, the cell surface changes from reflective to adaptive, allowing the light to penetrate the cell. The other parts of the cell remain reflective increasing the retention of the absorbed light within the cell.

In 2014 a system that combined an adaptive surface with a glass substrate that redirect the absorbed to a light absorber on the edges of the sheet. The system also included an array of fixed lenses/mirrors to concentrate light onto the adaptive surface. As the day continues, the concentrated light moves along the surface of the cell. That surface switches from reflective to adaptive when the light is most concentrated and back to reflective after the light moves along.

Manufacture

Solar cells share some of the same processing and manufacturing techniques as other semiconductor devices. However, the stringent requirements for cleanliness and quality control of semiconductor fabrication are more relaxed for solar cells, lowering costs.

Polycrystalline silicon wafers are made by wire-sawing block-cast silicon ingots into 180 to 350 micrometer wafers. The wafers are usually lightly p-type-doped. A surface diffusion of n-type dopants is performed on the front side of the wafer. This forms a p–n junction a few hundred nanometers below the surface.

Early solar-powered calculator

Anti-reflection coatings are then typically applied to increase the amount of light coupled into the solar cell. Silicon nitride has gradually replaced titanium dioxide as the preferred material, because of its excellent surface passivation qualities. It prevents carrier recombination at the cell surface. A layer several hundred nanometers thick is applied using PECVD. Some solar cells have textured front surfaces that, like anti-reflection coatings, increase the amount of light reaching the wafer. Such surfaces were first applied to single-crystal silicon, followed by multicrystalline silicon somewhat later.

A full area metal contact is made on the back surface, and a grid-like metal contact made up of fine "fingers" and larger "bus bars" are screen-printed onto the front surface using a silver paste. This is an evolution of the so-called "wet" process for applying electrodes, first described in a US patent filed in 1981 by Bayer AG. The rear contact is formed by screen-printing a metal paste, typically aluminium. Usually this contact covers the entire rear, though some designs employ a grid pattern. The paste is then fired at several hundred degrees Celsius to form metal electrodes in ohmic contact with the silicon. Some companies use an additional electro-plating step to increase efficiency. After the metal contacts are made, the solar cells are interconnected by flat wires or metal ribbons, and assembled into modules or "solar panels". Solar panels have a sheet of tempered glass on the front, and a polymer encapsulation on the back.

Manufacturers and Certification

National Renewable Energy Laboratory tests and validates solar technologies. Three reliable groups certify solar equipment: UL and IEEE (both U.S. standards) and IEC.

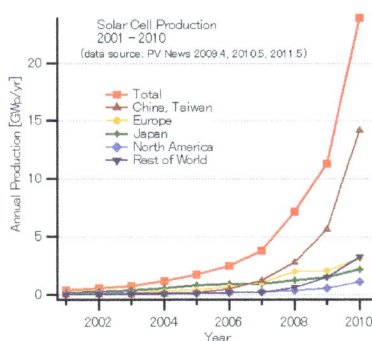

Solar cell production by region

Solar cells are manufactured in volume in Japan, Germany, China, Taiwan, Malaysia and the United States, whereas Europe, China, the U.S., and Japan have dominated (94% or more as of 2013) in installed systems. Other nations are acquiring significant solar cell production capacity.

Global PV cell/module production increased by 10% in 2012 despite a 9% decline in solar energy investments according to the annual "PV Status Report" released by the European Commission's Joint Research Centre. Between 2009 and 2013 cell production has quadrupled.

China

Due to heavy government investment, China has become the dominant force in solar cell manufacturing. Chinese companies produced solar cells/modules with a capacity of ~23 GW in 2013 (60% of global production).

Malaysia

In 2014, Malaysia was the world's third largest manufacturer of photovoltaics equipment, behind China and the European Union.

United States

Solar cell production in the U.S. has suffered due to the global financial crisis, but recovered partly due to the falling price of quality silicon.

Thin Film

A thin film is a layer of material ranging from fractions of a nanometer (monolayer) to several micrometers in thickness. The controlled synthesis of materials as thin films (a process referred to as deposition) is a fundamental step in many applications. A familiar example is the household mirror, which typically has a thin metal coating on the back of a sheet of glass to form a reflective interface. The process of silvering was once commonly used to produce mirrors, while more recently the metal layer is deposited using techniques such as sputtering. Advances in thin film deposition techniques during the 20th century have enabled a wide range of technological breakthroughs in areas such as magnetic recording media, electronic semiconductor devices, LEDs, optical coatings (such as antireflective coatings), hard coatings on cutting tools, and for both energy generation (e.g. thin film solar cells) and storage (thin-film batteries). It is also being applied to pharmaceuticals, via thin-film drug delivery.

In addition to their applied interest, thin films play an important role in the development and study of materials with new and unique properties. Examples include multiferroic materials, and superlattices that allow the study of quantum confinement by creating two-dimensional electron states.

Deposition

The act of applying a thin film to a surface is *thin-film deposition* – any technique for depositing a thin film of material onto a substrate or onto previously deposited layers. "Thin" is a relative term, but most deposition techniques control layer thickness within a few tens of nanometres. Molecular beam epitaxy and atomic layer deposition allow a single layer of atoms to be deposited at a time.

It is useful in the manufacture of optics (for reflective, anti-reflective coatings or self-cleaning glass, for instance), electronics (layers of insulators, semiconductors, and conductors form integrated circuits), packaging (i.e., aluminium-coated PET film), and in contemporary art. Similar processes are sometimes used where thickness is not important: for instance, the purification of copper by electroplating, and the deposition of silicon and enriched uranium by a CVD-like process after gas-phase processing.

Deposition techniques fall into two broad categories, depending on whether the process is primarily chemical or physical.

Chemical Deposition

Here, a fluid precursor undergoes a chemical change at a solid surface, leaving a solid layer. An everyday example is the formation of soot on a cool object when it is placed inside a flame. Since the fluid surrounds the solid object, deposition happens on every surface, with little regard to direction; thin films from chemical deposition techniques tend to be *conformal*, rather than *directional*.

Chemical deposition is further categorized by the phase of the precursor:

Plating relies on liquid precursors, often a solution of water with a salt of the metal to be deposited. Some plating processes are driven entirely by reagents in the solution (usually for noble metals), but by far the most commercially important process is electroplating. It was not commonly used in semiconductor processing for many years, but has seen a resurgence with more widespread use of chemical-mechanical polishing techniques.

Chemical solution deposition (CSD) or chemical bath deposition (CBD) uses a liquid precursor, usually a solution of organometallic powders dissolved in an organic solvent. This is a relatively inexpensive, simple thin film process that is able to produce stoichiometrically accurate crystalline phases. This technique is also known as the sol-gel method because the 'sol' (or solution) gradually evolves towards the formation of a gel-like diphasic system.

Spin coating or spin casting, uses a liquid precursor, or sol-gel precursor deposited onto a smooth, flat substrate which is subsequently spun at a high velocity to centrifugally spread the solution over the substrate. The speed at which the solution is spun and the viscosity of the sol determine the ultimate thickness of the deposited film. Repeated depositions can be carried out to increase the thickness of films as desired. Thermal treatment is often carried out in order to crystallize the amorphous spin coated film. Such crystalline films can exhibit certain preferred orientations after crystallization on single crystal substrates.

Chemical vapor deposition (CVD) generally uses a gas-phase precursor, often a halide or hydride of the element to be deposited. In the case of MOCVD, an organometallic gas is used. Commercial techniques often use very low pressures of precursor gas.

Plasma enhanced CVD (PECVD) uses an ionized vapor, or plasma, as a precursor. Unlike the soot example above, commercial PECVD relies on electromagnetic means (electric current, microwave excitation), rather than a chemical reaction, to produce a plasma.

Atomic layer deposition (ALD) uses gaseous precursor to deposit conformal thin films one layer at a time. The process is split up into two half reactions, run in sequence and repeated for each layer, in order to ensure total layer saturation before beginning the next layer. Therefore, one reactant is deposited first, and then the second reactant is deposited, during which a chemical reaction occurs on the substrate, forming the desired composition. As a result of the stepwise, the process is slower than CVD, however it can be run at low temperatures, unlike CVD.

Physical Deposition

Physical deposition uses mechanical, electromechanical or thermodynamic means to produce a thin film of solid. An everyday example is the formation of frost. Since most engineering materials are held together by relatively high energies, and chemical reactions are not used to store these energies, commercial physical deposition systems tend to require a low-pressure vapor environment to function properly; most can be classified as physical vapor deposition (PVD).

The material to be deposited is placed in an energetic, entropic environment, so that particles of material escape its surface. Facing this source is a cooler surface which draws energy from these particles as they arrive, allowing them to form a solid layer. The whole system is kept in a vacuum deposition chamber, to allow the particles to travel as freely as possible. Since particles tend to follow a straight path, films deposited by physical means are commonly *directional*, rather than *conformal*.

Examples of physical deposition include: A thermal evaporator that uses an electric resistance heater to melt the material and raise its vapor pressure to a useful range. This is done in a high vacuum, both to allow the vapor to reach the substrate without reacting with or scattering against other gas-phase atoms in the chamber, and reduce the incorporation of impurities from the residual gas in the vacuum chamber. Obviously, only materials with a much higher vapor pressure than the heating element can be deposited without contamination of the film. Molecular beam epitaxy is a particularly sophisticated form of thermal evaporation.

An electron beam evaporator fires a high-energy beam from an electron gun to boil a small spot of material; since the heating is not uniform, lower vapor pressure materials can be deposited. The beam is usually bent through an angle of 270° in order to ensure that the gun filament is not directly exposed to the evaporant flux. Typical deposition rates for electron beam evaporation range from 1 to 10 nanometres per second.

In molecular beam epitaxy (MBE), slow streams of an element can be directed at the substrate, so that material deposits one atomic layer at a time. Compounds such as gallium arsenide are usually deposited by repeatedly applying a layer of one element (i.e., gallium), then a layer of the other (i.e., arsenic), so that the process is chemical, as well as physical. The beam of material can be generated by either physical means (that is, by a furnace) or by a chemical reaction (chemical beam epitaxy).

Sputtering relies on a plasma (usually a noble gas, such as argon) to knock material from a "target" a few atoms at a time. The target can be kept at a relatively low temperature, since the process is not one of evaporation, making this one of the most flexible deposition techniques. It is especially useful for compounds or mixtures, where different components would otherwise tend to evaporate at different rates. Note, sputtering's step coverage is more or less conformal. It is also widely used in the optical media. The manufacturing of all formats of CD, DVD, and BD are done with the help of this technique. It is a fast technique and also it provides a good thickness control. Presently, nitrogen and oxygen gases are also being used in sputtering.

Pulsed laser deposition systems work by an ablation process. Pulses of focused laser light vaporize the surface of the target material and convert it to plasma; this plasma usually reverts to a gas before it reaches the substrate.

Cathodic arc deposition (arc-PVD) which is a kind of ion beam deposition where an electrical arc is created that literally blasts ions from the cathode. The arc has an extremely high power density resulting in a high level of ionization (30–100%), multiply charged ions, neutral particles, clusters and macro-particles (droplets). If a reactive gas is introduced during the evaporation process, dissociation, ionization and excitation can occur during interaction with the ion flux and a compound film will be deposited.

Electrohydrodynamic deposition (electrospray deposition) is a relatively new process of thin film deposition. The liquid to be deposited, either in the form of nano-particle solution or simply a solution, is fed to a small capillary nozzle (usually metallic) which is connected to a high voltage. The substrate on which the film has to be deposited is connected to ground. Through the influence of electric field, the liquid coming out of the nozzle takes a conical shape (Taylor cone) and at the apex of the cone a thin jet emanates which disintegrates into very fine and small positively charged droplets under the influence of Rayleigh charge limit. The droplets keep getting smaller and smaller and ultimately get deposited on the substrate as a uniform thin layer.

Growth Modes

Frank-van-der-Merwe ("layer-by-layer"). In this growth mode the adsorbate-surface and adsorbate-adsorbate interactions are balanced. This type of growth requires lattice matching, and hence considered an "ideal" growth mechanism.

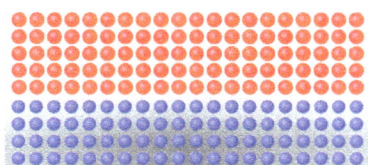

Frank-van-der-Merwe mode

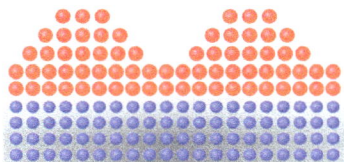

Stranski-Krastanow mode

Stranski–Krastanov growth ("joint islands" or "layer-plus-island"). In this growth mode the adsorbate-surface interactions are stronger than adsorbate-adsorbate interactions.

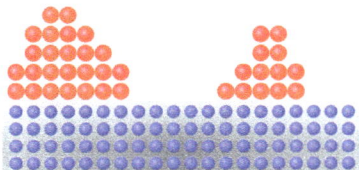

Volmer-Weber mode

Volmer-Weber ("isolated islands"). In this growth mode the adsorbate-adsorbate interactions are stronger than adsorbate-surface interactions, hence "islands" are formed right away.

Epitaxy

A subset of thin film deposition processes and applications is focused on the so-called epitaxial growth of materials, the deposition of crystalline thin films that grow following the crystalline structure of the substrate. The term epitaxy comes from the Greek roots epi, meaning "above", and taxis, meaning "an ordered manner". It can be translated as "arranging upon".

The term homoepitaxy refers to the specific case in which a film of the same material is grown on a crystalline substrate. This technology is used, for instance, to grow a film which is more pure than the substrate, has a lower density of defects, and to fabricate layers having different doping levels. Heteroepitaxy refers to the case in which the film being deposited is different than the substrate.

Techniques used for epitaxial growth of thin films include molecular beam epitaxy, chemical vapor deposition, and pulsed laser deposition.

Thin-film Photovoltaic Cells

Thin-film technologies are also being developed as a means of substantially reducing the cost of solar cells. The rationale for this is thin film solar cells are cheaper to manufacture owing to their reduced material costs, energy costs, handling costs and capital costs. This is especially represented in the use of printed electronics (roll-to-roll) processes. Other thin-film technologies, that are still in an early stage of ongoing research or with limited commercial availability, are often classified as emerging or third generation photovoltaic cells and include, organic, dye-sensitized, and polymer solar cells, as well as quantum dot, copper zinc tin sulfide, nanocrystal and perovskite solar cells.

Thin-film Batteries

Thin-film printing technology is being used to apply solid-state lithium polymers to a variety of substrates to create unique batteries for specialized applications. Thin-film batteries can be deposited directly onto chips or chip packages in any shape or size. Flexible batteries can be made by printing onto plastic, thin metal foil, or paper.

Thin-film Nanocomposites

Vertically aligned heteroepitaxial nanocomposite thin films have two phases grown epitaxially on given substrates and form unqiue nanostructures like nanowires in matrix. This kind of nanostructure has very large vertical interfacial area and tunable vertical strain, which can be used to tune functional properties. The STEM image shows a plan-view look of MgO nanowires (dark contrast) grown in La0.7Sr0.3MnO3 thin film matrix (white contrast).

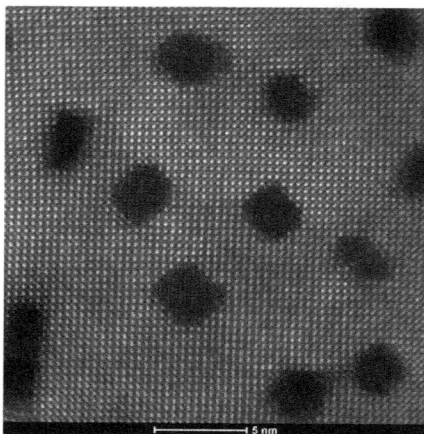

Top view of MgO nanowires grown in La0.7Sr0.3MnO3 thin film matrix

Carbon Nanotube Metal Matrix Composites

Carbon nanotube metal matrix composites (CNT-MMC) are an emerging class of new materials that are being developed to take advantage of the high tensile strength and electrical conductivity of carbon nanotube materials. Critical to the realization of CNT-MMC possessing optimal properties in these areas are the development of synthetic techniques that are (a) economically producible, (b) provide for a homogeneous dispersion of nanotubes in the metallic matrix, and (c) lead to strong interfacial adhesion between the metallic matrix and the carbon nanotubes. Since the development of CNT-MMC is still in the research phase, the current focus is primarily on improving these latter two areas.

Carbon Nanotubes Reinforced Metal Matrix Composites Production Methods

According to the new production systems, Carbon nanotubes reinforced metal matrix composites (CNT-MMC) may produce several different methods. These production methods are:

Powder metallurgy Route Technics

1. Conventional Sintering
2. Hot Pressing
3. Spark Plasma Sintering
4. Deformation Processing
5. Hot Extrusion
6. Semi-solid Powder Processing

Electrochemical Routes(for Non-structural Applications)

1. Electro-deposition
2. Electroless Deposition

Thermal Spraying

1. Plasma Spraying
2. HVOF Spraying
3. Cold Kinetic Spraying

Melt Processing

1. Casting
2. Melt Infiltration

Novel Techniques

1. Molecular Level Mixing

2. Sputtering

3. Sandwich Processing

4. Torsion/Friction Processing

5. CVD and PVD (Physical vapor deposition)

6. Nanoscale Dispersion

7. Pulsed laser deposition

Indigenous Techniques

1. Such as molecular level mixing (in which CNTs are dispersed into a metal-salt bath, forming a metal-CNT precursor).

Powder Metallurgy Techniques

Sintering is one of the oldest method in production technics and used to produce density-controlled materials and components from metal or ceramic powders by applying thermal energy. Synthesis and sintering of nanocrystalline ceramic powders have attracted much attention due to their promising properties. The high active surface area of nanopowders results in lowering sintering temperature relative to coarser powders. Although low temperature sintering suppresses the grain growth, high density of interfaces and grain boundaries in nanocrystalline powders leads to accelerated grain growth during sintering.

1. Conventional Sintering is the simplest method for producing CNT metal matrix composite compacts.The CNTs and metal powders are mixed by a process of mechanical alloying/ blending and then are compressed to form a green compact, which is then sintered to get the final product. Metallic compacts are subject to oxidation as compared to ceramics and hence the sintering has to be done in an inert atmosphere or under vacuum. One major drawback of this processing route is the inability to tailor the CNT distribution within the metallic matrix.

2. Microwave sintering is one of them and fundamentally different from conventional sintering.In microwave sintering process,the material is heated internally and volumetrically unlike in a conventional process where heat originates from an external heating source. Sintering cycle time for microwave sintering is much shorter as compared with the conventional sintering cycle.

3. Spark plasma sintering is a quite new technique which takes only a few minutes to complete a sintering process compared to conventional sintering which may take hours or even days for the same.High sintering rate is possible in SPS since high heating rates can be easily attained due to internal heating of the sample as oppsed to external heating seen in case of conventional sintering.For conventional sintering usually a green compact needs to be

prepared externally using a suitable die and hydraulic machine for applying the necessary pressure.In SPS the powder is directly fed into the graphite dies and the die is enclosed with suitable punches.All types of materials, even those difficult to densify can be easily sintered in SPS.Due to advantage of high heating rate and less holding time, SPS can restrict the unwanted sintering reactions in highly reactive systems as opposed to conventional sintering and hence formation of undesirable product phases can be avoided.

4. Semi-solid Powder Processing (SPP) is a unique method that fabricates composites materials with powder mixtures in the semi-solid states. Starting with metal-CNT powder mixture, the metal powder is heated to the semi-solid state, and pressure is applied to form the metal matrix composites. This method features many advantages such as simple and fast process and flexible property tailoring.

Carbon nanotube dispersing and CNT breakage during mixing

One common method to disperse the CNT into the metal matrix is mechanical alloying. However, many researchers reported the length reduction and damage of CNTs during mechanical alloying process.

Mechanical Properties

Carbon nanotubes are the strongest and stiffest materials yet discovered in terms of tensile strength and elastic modulus respectively. This strength results from the covalent sp^2 bonds formed between the individual carbon atoms.Multi-walled carbon nanotube was tested to have a tensile strength of 63 gigapascals (GPa). Further studies, conducted in 2008, revealed that individual CNT shells have strengths of up to ~100 GPa, which is in good agreement with quantum/atomistic models. Since carbon nanotubes have a low density for a solid of 1.3 to 1.4 g/cm^3, its specific strength of up to 48,000 $kN \cdot m \cdot kg^{-1}$ is the best of known materials, compared to high-carbon steel's 154 $kN \cdot m \cdot kg^{-1}$. CNTs are not nearly as strong under compression. Because of their hollow structure and high aspect ratio, they tend to undergo buckling when placed under compressive, torsional, or bending stress.

Comparison of mechanical properties			
Material	Young's modulus (TPa)	Tensile strength (GPa)	Elongation at break (%)
SWNT[E]	~1 (from 1 to 5)	13–53	16
Armchair SWNT[T]	0.94	126.2	23.1
Zigzag SWNT[T]	0.94	94.5	15.6–17.5
Chiral SWNT	0.92		
MWNT[E]	0.2–0.8–0.95	11–63–150	
Stainless steel[E]	0.186–0.214	0.38–1.55	15–50
Kevlar–29&149[E]	0.06–0.18	3.6–3.8	~2

[E]Experimental observation; [T]Theoretical prediction

Potential Applications

The exhibit 'future soldier', designed for US Army

Type 10 MBT consist of Nano-crystal steel (or Triple Hardness Steel), Modular ceramic composite armor, Light weight upper armor.

Leopard 2SG of the Singapore Army upgraded with AMAP Composite Armour by IBD & ST Kinetics

Nanonetwork

Nanonetworks are expected to expand the capabilities of single nanomachines both in terms of complexity and range of operation by allowing them to coordinate, share and fuse information. CNT metal matrix composites enable new applications of nanotechnology in the military technology and industrial and consumer goods applications.

Nanorobotics

Nanomachines are largely in the research-and-development phase, but some primitive molecular machines have been tested. An example is a sensor having a switch approximately 1.5 nanometers across, capable of counting specific molecules in a chemical sample. The first useful applications of nanomachines might be in medical technology, which could be used to identify and destroy cancer cells. Another potential application is the detection of toxic chemicals, and the measurement of their concentrations, in the environment.In addition CNT-MM composite will be the main material for the military robots, especially to strength the robot soldier's armours.

Future Soldier

Today's militaries often use high-quality helmets made of ballistic materials such as Kevlar and Aramid, which offer improved protection. Some helmets also have good non-ballistic protective qualities, though many do not. Non-ballistic injuries may be caused by many things, such as concussive shock waves from explosions, physical attacks, motor vehicle accidents, or falls. Another application for the future soldier is powered exoskeleton system. Powered exoskeleton, also known as powered armor, or exoframe, is a powered mobile machine consisting primarily of an exoskeleton-like framework worn by a person and a power supply that supplies at least part of the activation-energy for limb movement.Powered exoskeletons are designed to assist and protect the soldiers and officers.Currently MIT is working on combat jackets that use CNT fibers to stop bullets and to monitor the condition of the wearer.

Advanced Modular Armor Protection

Advanced Modular Armor Protection (AMAP) is modular composite armour concept, developed by the German company IBD Deisenroth Engineering. According to IBD AMAP is a 4th generation composite armour, making use of nano-ceramics and modern steel alloy technologies. AMAP is making use of new advanced steel alloys, Aluminium-Titanium alloys, nanometric steels, ceramics and nano-ceramics. The new high-hardened steel needs 30% less thickness to offer the same protection level as ARMOX500Z High Hard Armour steel. While Titanium requires only 58% as much weight as rolled homogeneous armour (RHA) for reaching the same level of protection, *Mat 7720 new*, a newly developed Aluminium-Titanium alloy, needs only 38% of the weight. That means that this alloy is more than twice as protective as RHA of the same weight.

AMAP is also making use of new nano-ceramics, which are harder and lighter than current ceramics, while having multi-hit capability. Normal ceramic tiles and a liner backing have a mass-efficiency (E_M) value of 3 compared to normal steel armour, while it fulfills STANAG 4569. The new nano-crystalline ceramic materials should increase the hardness compared to current ceramics by 70% and the weight reduction is 30%, therefore the E_M value is larger than 4. Furthermore, the higher fracture toughness increases the general multi-hit capability. Some AMAP-modules might consist of this new ceramic tiles glued on a backing liner and overlaid by a cover, a concept which is also used by MEXAS. Lightweight SLAT armour is also part of the AMAP family.

Nano Armour

The TK-X (MBT-X) project, the new Type 10 main battle tank design use composition of modular components of nano-crystal steel (nor Triple Hardness Steel), Modular ceramic composite armor, partly reinforced MMC and Light weight upper armor.

Materiomics

Materiomics is defined as the study of the material properties of natural and synthetic materials by examining fundamental links between processes, structures and properties at multiple scales, from nano to macro, by using systematic experimental, theoretical or computational methods and refers to the study of the processes, structures and properties of materials from a fundamental, systematic perspective by incorporating all relevant scales, from nano to macro, in the synthesis and function of materials and structures. The integrated view of these interactions at all scales is referred to as a material's materiome.

Materiomics includes the study of a broad range of materials, which includes metals, ceramics and polymers as well as biological materials and tissues and their interaction with synthetic materials. Materiomics finds applications in elucidating the biological role of materials in biology, for instance in the progression and diagnosis or the treatment of diseases. Others have proposed to apply materiomics concepts to help identify new material platforms for tissue engineering applications, for instance for the de novo development of biomaterials. Materiomics might also hold promises for nanoscience and nanotechnology, where the understanding of material concepts at multiple scales could enable the bottom-up development of new structures and materials or devices, including biomimetic and bioinspired structures.

Nanotough

Nanotough is to obtain a deeper understanding of the interfacial structure of nanocomposites within a polyolefin matrix and thus use nanoparticles like nanoclay to turn upside down the construction of a number of well-known products, where today metals or plastics are used in for example cars or aircraft. The project will enable realization of the great performance potential of these materials through development of novel multiphase and hybrid nanocomposites.

Nanotough project aims to improve the stiffness of polyolefin nanocomposites while not only maintaining but also improving the toughness of the matrix considerably. The technical objective is to optimize and, through novel interface design, to develop new cost effective hybrid (nanofiller-fibre) nanocomposites as an alternative to heavily filled polymers and expensive engineering polymers and fulfill industry requirements for high performance materials in high tech applications.

Exfoliated Graphite Nanoplatelets

Exfoliated graphite nano-platelets (xGnP) are new types of nanoparticles made from graphite. These nanoparticles consist of small stacks of graphene that are 1 to 15 nanometers thick, with diameters ranging from sub-micrometre to 100 micrometres. The X-ray diffractogram of this material would resemble that of graphite, in that the 002 peak would still appear at ~26° 2 theta. However, the peak would appear considerably smaller and broader. These features indicate that the interplanar distance in exfoliated graphite is similar to that of the parent graphite, but the stack size (of graphene layers) is small. Since xGnP is composed of the same material as carbon nanotubes, it shares many of the electrochemical characteristics, although not the tensile strength. The platelet shape, however, offers xGnP edges that are easier to modify chemically for enhanced dispersion in polymers.

Composite materials made with polymers, like plastics, nylon, or rubber, can be made electrically or thermally conductive with the addition of small amounts of xGnP. These nanoparticles can change the fundamental properties of plastics, enabling them to perform more like metals with metallic properties. These new nanoparticles also improve barrier properties, modulus, and surface toughness when used in composites.

Properties

Graphene is extremely electrically conductive material. In turn, xGnP has a percolation threshold for conductivity of 1.9 wt% in thermoplastic matrix.At densities of 2–5 wt%, conductivity reaches sufficient levels to provide electromagnetic shielding. xGnP can also be combined with glass fibers or other matrix materials to provide sufficient conductivity for electrostatic painting or other applications requiring electrical conductivity.

xGnP significantly outperforms most other forms of carbon in terms of thermal conductivity when used at densities of 20 wt% in control resins. At these densities, xGnP also confers significant electrical conductivity as well as improved mechanical properties to most thermoplastic, thermoset, or elastomeric systems. At lesser densities, xGnP adds thermal stability to a variety of matrix materials.

As opposed to materials like carbon black, xGnP improves mechanical properties of most composites, particularly stiffness and tensile strength. Elastomeric compounds have been shown to experience increased life and reduced surface wear when reinforced with xGnP.

Because of the platelet shape, xGnP significantly improves the impermeability of composites when used at densities of ~3 wt% or greater. xGnP particles can be aligned using electric field, although alignment is not necessary for use in most extrusion systems. Because xGnP also imparts electrical conductivity at these densities, the resulting composites offer attractive cost savings for applications like fuel lines or fuel tank linings.

Colloid

A colloid, in chemistry, is a mixture in which one substance of microscopically dispersed insoluble particles is suspended throughout another substance. Sometimes the dispersed substance alone is called the colloid; the term colloidal suspension refers unambiguously to the overall mixture (although a narrower sense of the word *suspension* is distinguished from colloids by larger particle size). Unlike a solution, whose solute and solvent constitute only one phase, a colloid has a dispersed phase (the suspended particles) and a continuous phase (the medium of suspension). To qualify as a colloid, the mixture must be one that does not settle or would take a very long time to settle appreciably.

The dispersed-phase particles have a diameter between approximately 1 and 1000 nanometers. Such particles are normally easily visible in an optical microscope, although at the smaller size range (r<250 nm), an ultramicroscope or an electron microscope may be required. Homogeneous mixtures with a dispersed phase in this size range may be called *colloidal aerosols*, *colloidal emulsions*, *colloidal foams*, *colloidal dispersions*, or *hydrosols*. The dispersed-phase particles or droplets are affected largely by the surface chemistry present in the colloid.

Milk is an emulsified colloid of liquid butterfat globules dispersed within a water-based solution.

Some colloids are translucent because of the Tyndall effect, which is the scattering of light by particles in the colloid. Other colloids may be opaque or have a slight color.

Colloidal suspensions are the subject of interface and colloid science. This field of study was introduced in 1861 by Scottish scientist Thomas Graham.

IUPAC Definition

Colloidal: State of subdivision such that the molecules or polymolecular particles dispersed in a medium have at least one dimension between approximately 1 nm and 1 μm, or that in a system discontinuities are found at distances of that order.

Classification

Because the size of the dispersed phase may be difficult to measure, and because colloids have the appearance of solutions, colloids are sometimes identified and characterized by their physico-chemical and transport properties. For example, if a colloid consists of a solid phase dispersed in a liquid, the solid particles will not diffuse through a membrane, whereas with a true solution the dissolved ions or molecules will diffuse through a membrane. Because of the size exclusion, the colloidal particles are unable to pass through the pores of an ultrafiltration membrane with a size smaller than their own dimension. The smaller the size of the pore of the ultrafiltration membrane, the lower the concentration of the dispersed colloidal particles remaining in the ultrafiltered liquid. The measured value of the concentration of a truly dissolved species will thus depend on the experimental conditions applied to separate it from the colloidal particles also dispersed in the liquid. This is particularly important for solubility studies of readily hydrolyzed species such as Al, Eu, Am, Cm, or organic matter complexing these species. Colloids can be classified as follows:

Medium/phase		Dispersed phase		
Gas		Liquid	Solid	
Dispersion medium	Gas	None All gases are miscible and thus do not form colloids	Liquid aerosol Examples: fog, hair sprays	Solid aerosol Examples: smoke, ice cloud, atmospheric particulate matter
	Liquid	Foam Example: whipped cream, shaving cream	Emulsion Examples: milk, mayonnaise, hand cream; latex	Sol Examples: pigmented ink, blood
	Solid	Solid foam Examples: aerogel, styrofoam, pumice	Gel Examples: agar, gelatin, jelly	Solid sol Example: cranberry glass

Based on the nature of interaction between the dispersed phase and the dispersion medium, colloids can be classified as: Hydrophilic colloids: These are water-loving colloids. The colloid particles are attracted toward water. They are also called reversible sols. Hydrophobic colloids: These are opposite in nature to hydrophilic colloids. The colloid particles are repelled by water. They are also called irreversible sols.

In some cases, a colloid suspension can be considered a homogeneous mixture. This is because the distinction between "dissolved" and "particulate" matter can be sometimes a matter of approach, which affects whether or not it is homogeneous or heterogeneous.

Hydrocolloids

A *hydrocolloid* is defined as a colloid system wherein the colloid particles are hydrophilic polymers dispersed in water. A hydrocolloid has colloid particles spread throughout water, and depending on the quantity of water available that can take place in different states, e.g., gel or sol (liquid). Hydrocolloids can be either irreversible (single-state) or reversible. For example, agar, a reversible hydrocolloid of seaweed extract, can exist in a gel and solid state, and alternate between states with the addition or elimination of heat.

Many hydrocolloids are derived from natural sources. For example, agar-agar and carrageenan are extracted from seaweed, gelatin is produced by hydrolysis of proteins of mammalian and fish origins, and pectin is extracted from citrus peel and apple pomace.

Gelatin desserts like jelly or Jell-O are made from gelatin powder, another effective hydrocolloid. Hydrocolloids are employed in food mainly to influence texture or viscosity (e.g., a sauce). Hydrocolloid-based medical dressings are used for skin and wound treatment.

Other main hydrocolloids are xanthan gum, gum arabic, guar gum, locust bean gum, cellulose derivatives as carboxymethyl cellulose, alginate and starch.

Interaction Between Particles

The following forces play an important role in the interaction of colloid particles:

- Excluded volume repulsion: This refers to the impossibility of any overlap between hard particles.

- Electrostatic interaction: Colloidal particles often carry an electrical charge and therefore attract or repel each other. The charge of both the continuous and the dispersed phase, as well as the mobility of the phases are factors affecting this interaction.

- van der Waals forces: This is due to interaction between two dipoles that are either permanent or induced. Even if the particles do not have a permanent dipole, fluctuations of the electron density gives rise to a temporary dipole in a particle. This temporary dipole induces a dipole in particles nearby. The temporary dipole and the induced dipoles are then attracted to each other. This is known as van der Waals force, and is always present (unless the refractive indexes of the dispersed and continuous phases are matched), is short-range, and is attractive.

- Entropic forces: According to the second law of thermodynamics, a system progresses to a state in which entropy is maximized. This can result in effective forces even between hard spheres.

- Steric forces between polymer-covered surfaces or in solutions containing non-adsorbing polymer can modulate interparticle forces, producing an additional steric repulsive force (which is predominantly entropic in origin) or an attractive depletion force between them. Such an effect is specifically searched for with tailor-made superplasticizers developed to increase the workability of concrete and to reduce its water content.

Preparation

There are two principal ways of preparation of colloids:

- Dispersion of large particles or droplets to the colloidal dimensions by milling, spraying, or application of shear (e.g., shaking, mixing, or high shear mixing).

- Condensation of small dissolved molecules into larger colloidal particles by precipitation, condensation, or redox reactions. Such processes are used in the preparation of colloidal silica or gold.

Stabilization (Peptization)

The stability of a colloidal system is defined by particles remaining suspended in solution at equilibrium.

Stability is hindered by aggregation and sedimentation phenomena, which are driven by the colloid's tendency to reduce surface energy. Reducing the interfacial tension will stabilize the colloidal system by reducing this driving force.

Aggregation is due to the sum of the interaction forces between particles. If attractive forces (such as van der Waals forces) prevail over the repulsive ones (such as the electrostatic ones) particles aggregate in clusters.

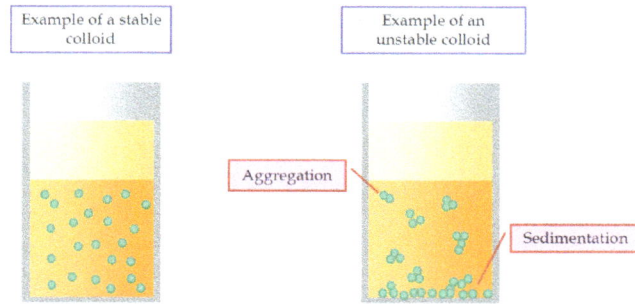

Examples of a stable and of an unstable colloidal dispersion.

Electrostatic stabilization and steric stabilization are the two main mechanisms for stabilization against aggregation.

- Electrostatic stabilization is based on the mutual repulsion of like electrical charges. In general, different phases have different charge affinities, so that an electrical double layer forms at any interface. Small particle sizes lead to enormous surface areas, and this effect is greatly amplified in colloids. In a stable colloid, mass of a dispersed phase is so low that its buoyancy or kinetic energy is too weak to overcome the electrostatic repulsion between charged layers of the dispersing phase.

- Steric stabilization consists in covering the particles in polymers which prevents the particle to get close in the range of attractive forces.

A combination of the two mechanisms is also possible (electrosteric stabilization). All the above-mentioned mechanisms for minimizing particle aggregation rely on the enhancement of the repulsive interaction forces.

Electrostatic and steric stabilization do not directly address the sedimentation/floating problem.

Particle sedimentation (and also floating, although this phenomenon is less common) arises from a difference in the density of the dispersed and of the continuous phase. The higher the difference in densities, the faster the particle settling.

- The gel network stabilization represents the principal way to produce colloids stable to both aggregation and sedimentation.

The method consists in adding to the colloidal suspension a polymer able to form a gel network and characterized by shear thinning properties. Examples of such substances are xanthan and guar gum.

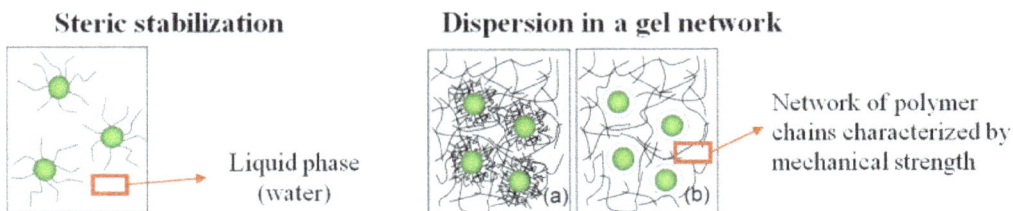

Steric and gel network stabilization.

Particle settling is hindered by the stiffness of the polymeric matrix where particles are trapped. In addition, the long polymeric chains can provide a steric or electrosteric stabilization to dispersed particles.

The rheological shear thinning properties find beneficial in the preparation of the suspensions and in their use, as the reduced viscosity at high shear rates facilitates deagglomeration, mixing and in general the flow of the suspensions.

Destabilization

Unstable colloidal dispersions can form flocs as the particles aggregate due to interparticle attractions. In this way photonic glasses can be grown. This can be accomplished by a number of different methods:

- Removal of the electrostatic barrier that prevents aggregation of the particles. This can be accomplished by the addition of salt to a suspension or changing the pH of a suspension to effectively neutralize or "screen" the surface charge of the particles in suspension. This removes the repulsive forces that keep colloidal particles separate and allows for coagulation due to van der Waals forces.

- Addition of a charged polymer flocculant. Polymer flocculants can bridge individual colloidal particles by attractive electrostatic interactions. For example, negatively charged colloidal silica or clay particles can be flocculated by the addition of a positively charged polymer.

- Addition of non-adsorbed polymers called depletants that cause aggregation due to entropic effects.

- Physical deformation of the particle (e.g., stretching) may increase the van der Waals forces more than stabilization forces (such as electrostatic), resulting coagulation of colloids at certain orientations.

Unstable colloidal suspensions of low-volume fraction form clustered liquid suspensions, wherein individual clusters of particles fall to the bottom of the suspension (or float to the top if the particles are less dense than the suspending medium) once the clusters are of sufficient size for the Brownian forces that work to keep the particles in suspension to be overcome by gravitational forces. However, colloidal suspensions of higher-volume fraction form colloidal gels with viscoelastic properties. Viscoelastic colloidal gels, such as bentonite and toothpaste, flow like liquids under shear, but maintain their shape when shear is removed. It is for this reason that toothpaste can be squeezed from a toothpaste tube, but stays on the toothbrush after it is applied.

Monitoring Stability

Multiple light scattering coupled with vertical scanning is the most widely used technique to monitor the dispersion state of a product, hence identifying and quantifying destabilisation phenomena. It works on concentrated dispersions without dilution. When light is sent through the sample, it is backscattered by the particles / droplets. The backscattering intensity is directly proportional to the size and volume fraction of the dispersed phase. Therefore, local changes in concentration (*e.g.*Creaming and Sedimentation) and global changes in size (*e.g.* flocculation, coalescence) are detected and monitored.

1 / Positionning
Positionnement

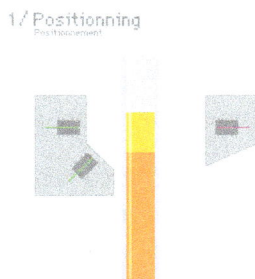

Measurement principle of multiple light scattering coupled with vertical scanning

Accelerating Methods for Shelf Life Prediction

The kinetic process of destabilisation can be rather long (up to several months or even years for some products) and it is often required for the formulator to use further accelerating methods in order to reach reasonable development time for new product design. Thermal methods are the most commonly used and consists in increasing temperature to accelerate destabilisation (below critical temperatures of phase inversion or chemical degradation). Temperature affects not only the viscosity, but also interfacial tension in the case of non-ionic surfactants or more generally interactions forces inside the system. Storing a dispersion at high temperatures enables to simulate real life conditions for a product (e.g. tube of sunscreen cream in a car in the summer), but also to accelerate destabilisation processes up to 200 times. Mechanical acceleration including vibration, centrifugation and agitation are sometimes used. They subject the product to different forces that pushes the particles / droplets against one another, hence helping in the film drainage. However, some emulsions would never coalesce in normal gravity, while they do under artificial gravity. Moreover, segregation of different populations of particles have been highlighted when using centrifugation and vibration.

As a Model System for Atoms

In physics, colloids are an interesting model system for atoms. Micrometre-scale colloidal particles are large enough to be observed by optical techniques such as confocal microscopy. Many of the forces that govern the structure and behavior of matter, such as excluded volume interactions or electrostatic forces, govern the structure and behavior of colloidal suspensions. For example, the same techniques used to model ideal gases can be applied to model the behavior of a hard sphere colloidal suspension. In addition, phase transitions in colloidal suspensions can be studied in real time using optical techniques, and are analogous to phase transitions in liquids. In many interesting cases optical fluidity is used to control colloid suspensions.

Crystals

A colloidal crystal is a highly ordered array of particles that can be formed over a very long range (typically on the order of a few millimeters to one centimeter) and that appear analogous to their atomic or molecular counterparts. One of the finest natural examples of this ordering phenomenon can be found in precious opal, in which brilliant regions of pure spectral color result from close-packed domains of amorphous colloidal spheres of silicon dioxide (or silica, SiO_2). These spherical particles precipitate in highly siliceous pools in Australia and elsewhere, and form these highly ordered arrays after years of sedimentation and compression under hydrostatic and gravitational forces. The periodic arrays of submicrometre spherical particles provide similar arrays of interstitial voids, which act as a natural diffraction grating for visible light waves, particularly when the interstitial spacing is of the same order of magnitude as the incident lightwave.

Thus, it has been known for many years that, due to repulsive Coulombic interactions, electrically charged macromolecules in an aqueous environment can exhibit long-range crystal-like correlations with interparticle separation distances, often being considerably greater than the individual particle diameter. In all of these cases in nature, the same brilliant iridescence (or play of colors) can be attributed to the diffraction and constructive interference of visible lightwaves that satisfy Bragg's law, in a matter analogous to the scattering of X-rays in crystalline solids.

The large number of experiments exploring the physics and chemistry of these so-called "colloidal crystals" has emerged as a result of the relatively simple methods that have evolved in the last 20 years for preparing synthetic monodisperse colloids (both polymer and mineral) and, through various mechanisms, implementing and preserving their long-range order formation.

In Biology

In the early 20th century, before enzymology was well understood, colloids were thought to be the key to the operation of enzymes; i.e., the addition of small quantities of an enzyme to a quantity of water would, in some fashion yet to be specified, subtly alter the properties of the water so that it would break down the enzyme's specific substrate,such as a solution of ATPase breaking down ATP. Furthermore, life itself was explainable in terms of the aggregate properties of all the colloidal substances that make up an organism. As more detailed knowledge of biology and biochemistry developed, the colloidal theory was replaced by the macromolecular theory, which explains enzymes as a collection of identical huge molecules that act as very tiny machines, freely moving about between the water molecules of the solution and individually operating on the substrate, no more mysterious than a factory full of machinery. The properties of the water in the solution are not altered, other than the simple osmotic changes that would be caused by the presence of any solute. In humans, both the thyroid gland and the intermediate lobe (*pars intermedia*) of the pituitary gland contain colloid follicles.

In The Environment

Colloidal particles can also serve as transport vector of diverse contaminants in the surface water (sea water, lakes, rivers, fresh water bodies) and in underground water circulating in fissured rocks (e.g. limestone, sandstone, granite). Radionuclides and heavy metals easily sorb onto colloids suspended in water. Various types of colloids are recognised: inorganic colloids (e.g. clay particles, silicates, iron oxy-hydroxides), organic colloids (humic and fulvic substances). When heavy metals or radionuclides form their own pure colloids, the term "*Eigencolloid*" is used to designate pure phases, i.e. pure $Tc(OH)_4$, $U(OH)_4$, or $Am(OH)_3$. Colloids have been suspected for the long-range transport of plutonium on the Nevada Nuclear Test Site. They have been the subject of detailed studies for many years. However, the mobility of inorganic colloids is very low in compacted bentonites and in deep clay formations because of the process of ultrafiltration occurring in dense clay membrane. The question is less clear for small organic colloids often mixed in porewater with truly dissolved organic molecules.

Intravenous Therapy

Colloid solutions used in intravenous therapy belong to a major group of volume expanders, and can be used for intravenous fluid replacement. Colloids preserve a high colloid osmotic pressure in the blood, and therefore, they should theoretically preferentially increase the intravascular volume, whereas other types of volume expanders called crystalloids also increase the interstitial volume and intracellular volume. However, there is still controversy to the actual difference in efficacy by this difference, and much of the research related to this use of colloids is based on fraudulent research by Joachim Boldt. Another difference is that crystalloids generally are much cheaper than colloids.

Gel

A gel is a solid jelly-like material that can have properties ranging from soft and weak to hard and tough. Gels are defined as a substantially dilute cross-linked system, which exhibits no flow when in the steady-state. By weight, gels are mostly liquid, yet they behave like solids due to a three-dimensional cross-linked network within the liquid. It is the crosslinking within the fluid that gives a gel its structure (hardness) and contributes to the adhesive stick (tack). In this way gels are a dispersion of molecules of a liquid within a solid in which the solid is the continuous phase and the liquid is the discontinuous phase. The word *gel* was coined by 19th-century Scottish chemist Thomas Graham by clipping from *gelatine*.

An upturned vial of hair gel

Composition

Gels consist of a solid three-dimensional network that spans the volume of a liquid medium and ensnares it through surface tension effects. This internal network structure may result from physical bonds (physical gels) or chemical bonds (chemical gels), as well as crystallites or other junctions that remain intact within the extending fluid. Virtually any fluid can be used as an extender including water (hydrogels), oil, and air (aerogel). Both by weight and volume, gels are mostly fluid in composition and thus exhibit densities similar to those of their constituent liquids. Edible jelly is a common example of a hydrogel and has approximately the density of water.

Polyionic Polymers

Polyionic polymers are polymers with an ionic functional group. The ionic charges prevent the formation of tightly coiled polymer chains. This allows them to contribute more to viscosity in their stretched state, because the stretched-out polymer takes up more space. See polyelectrolyte for more information.

Types

Hydrogels

A hydrogel is a network of polymer chains that are hydrophilic, sometimes found as a colloidal gel in which water is the dispersion medium. Hydrogels are highly absorbent (they can contain over 90% water) natural or synthetic polymeric networks. Hydrogels also possess a degree of flexibility very similar to natural tissue, due to their significant water content. The first appearance of the term 'hydrogel' in the literature was in 1894. Common uses for hydrogels include:

A micropump based on a hydrogel bar (4×0.3×0.05 mm size) actuated by applied voltage. This pump can be continuously operated with a 1.5 V battery for at least 6 months.

- Scaffolds in tissue engineering. When used as scaffolds, hydrogels may contain human cells to repair tissue. They mimic 3D microenvironment of cells.

- Hydrogel-coated wells have been used for cell culture

- Environmentally sensitive hydrogels (also known as 'Smart Gels' or 'Intelligent Gels'). These hydrogels have the ability to sense changes of pH, temperature, or the concentration of metabolite and release their load as result of such a change.

- Sustained-release drug delivery systems

- Providing absorption, desloughing and debriding of necrotic and fibrotic tissue

- Hydrogels that are responsive to specific molecules, such as glucose or antigens, can be used as biosensors, as well as in DDS.

- Disposable diapers where they absorb urine, or in sanitary napkins

- Contact lenses (silicone hydrogels, polyacrylamides, polymacon)

- EEG and ECG medical electrodes using hydrogels composed of cross-linked polymers (polyethylene oxide, polyAMPS and polyvinylpyrrolidone)

- Water gel explosives

- Rectal drug delivery and diagnosis

- Encapsulation of quantum dots

Other, less common uses include

- Breast implants

- Glue

- Granules for holding soil moisture in arid areas

- Dressings for healing of burn or other hard-to-heal wounds. Wound gels are excellent for helping to create or maintain a moist environment.

- Reservoirs in topical drug delivery; particularly ionic drugs, delivered by iontophoresis.

Common ingredients include polyvinyl alcohol, sodium polyacrylate, acrylate polymers and copolymers with an abundance of hydrophilic groups.

Natural hydrogel materials are being investigated for tissue engineering; these materials include agarose, methylcellulose, hyaluronan, and other naturally derived polymers.

Organogels

An organogel is a non-crystalline, non-glassy thermoreversible (thermoplastic) solid material composed of a liquid organic phase entrapped in a three-dimensionally cross-linked network. The liquid can be, for example, an organic solvent, mineral oil, or vegetable oil. The solubility and particle dimensions of the structurant are important characteristics for the elastic properties and firmness of the organogel. Often, these systems are based on self-assembly of the structurant molecules. (An example of formation of an undesired thermoreversible network is the occurrence of wax crystallization in petroleum.)

Organogels have potential for use in a number of applications, such as in pharmaceuticals, cosmetics, art conservation, and food.

Xerogels

A xerogel /zrodl/ is a solid formed from a gel by drying with unhindered shrinkage. Xerogels usually retain high porosity (15–50%) and enormous surface area (150–900 m^2/g), along with very small pore size (1–10 nm). When solvent removal occurs under supercritical conditions, the network does not shrink and a highly porous, low-density material known as an *aerogel* is produced. Heat treatment of a xerogel at elevated temperature produces viscous sintering (shrinkage of the xerogel due to a small amount of viscous flow) and effectively transforms the porous gel into a dense glass.

Nanocomposite Hydrogels

Nanocomposite hydrogels are also known as hybrid hydrogels, can be defined as highly hydrated polymeric networks, either physically or covalently crosslinked with each other and/or with nanoparticles or nanostructures. Nanocomposite hydrogels can mimic native tissue properties, structure and microenvironment due to their hydrated and interconnected porous structure. A wide range of nanoparticles, such as carbon-based, polymeric, ceramic, and metallic nanomaterials can be incorporated within the hydrogel structure to obtain nanocomposites with tailored functionality. Nanocomposite hydrogels can be engineered to possess superior physical, chemical, electrical, and biological properties.

Properties

Many gels display thixotropy – they become fluid when agitated, but resolidify when resting. In general, gels are apparently solid, jelly-like materials. By replacing the liquid with gas it is possible to prepare aerogels, materials with exceptional properties including very low density, high specific surface areas, and excellent thermal insulation properties.

Animal-produced Gels

Some species secrete gels that are effective in parasite control. For example, the long-finned pilot whale secretes an enzymatic gel that rests on the outer surface of this animal and helps prevent other organisms from establishing colonies on the surface of these whales' bodies.

Hydrogels existing naturally in the body include mucus, the vitreous humor of the eye, cartilage, tendons and blood clots. Their viscoelastic nature results in the soft tissue component of the body, disparate from the mineral-based hard tissue of the skeletal system. Researchers are actively developing synthetically derived tissue replacement technologies derived from hydrogels, for both temporary implants (degradable) and permanent implants (non-degradable). A review article on the subject discusses the use of hydrogels for nucleus pulposus replacement, cartilage replacement, and synthetic tissue models.

Applications

Many substances can form gels when a suitable thickener or gelling agent is added to their formula. This approach is common in manufacture of wide range of products, from foods to paints and adhesives.

In fiber optics communications, a soft gel resembling "hair gel" in viscosity is used to fill the plastic tubes containing the fibers. The main purpose of the gel is to prevent water intrusion if the buffer tube is breached, but the gel also buffers the fibers against mechanical damage when the tube is bent around corners during installation, or flexed. Additionally, the gel acts as a processing aid when the cable is being constructed, keeping the fibers central whilst the tube material is extruded around it.

Hydrogels for Cancer Therapy

Recently, Conde et al. developed a self-assembled dual-color RNA triple-helix structure comprising two miRNAs, a miR mimic (tumor suppressor miRNA) and an antagomiR (oncomiR inhibitor) provides outstanding capability to synergistically abrogate tumors. The authors hypothesized that efficacious delivery of the RNA triple-helix hydrogel nanoconjugates would be achieved by coating the breast tumor with the adhesive hydrogel scaffold that they have previously shown able to enhance the stability of embedded nanoparticles used for local gene and drug delivery using smart gold nanobeacons.

Conjugation of RNA triple-helices to dendrimers allows the formation of stable triplex nanoparticles, which form an RNA-triple-helix adhesive scaffold on interaction with dextran aldehyde, the latter able to chemically interact and adhere to natural tissue amines in the tumor. The authors also show that the self-assembled RNA-triple-helix conjugates remain functional in vitro and in vivo, and that they lead to nearly 90% levels of tumor shrinkage two weeks post-gel implantation in a triple-negative breast-cancer mouse model. These findings suggest that the RNA-triple-helix hydrogels can be used as an efficient anticancer platform to locally modulate the expression of endogenous miRs in cancer.

The authors reported on a novel strategy for concomitant oncomiR inhibition and tumor suppressor miR replacement therapy using a RNA triple-helix hydrogel scaffold that affords highly efficacious local anticancer therapy. Self-assembled RNA triple-helix conjugates remain functional in vitro with high selective uptake and control over miR expression compared to their respective single-stranded or double-stranded forms. Hence, cancer gene delivery systems should provide potent, selective and specific treatment to tumor cells only, unlike the standard delivery of most conventional chemotherapeutic drugs. This approach can be implemented to design self-assembled triplex structures from any other miR combination, or from other genetic materials including antisense DNA or siRNA to treat a range of diseases.

More recently, Conde et al. took one step-forward and reported that these hydrogels were also optimized and developed as prophylactic patches able to perform gene, chemo and phototherapy in a triple-combination approach to achieve complete tumor resection when applied to non-resected tumors and to the absence of tumor recurrence when applied following tumor resection. This study also identifies the molecular and genetic pathways triggered in response to the three therapeutic modalities – photo-, gene-, chemo-therapy – by tumor gene expression profiling in treated mice. Taken together the studies hydrogels are excellent candidates for a successful strategy of the proposed project and represent a rational treatment strategy following a comprehensive scrutiny of the tumor microenvironment and host response to different therapeutic modalities.

Copolymer

IUPAC Definition for Copolymer

A polymer derived from more than one species of monomer.

Note: Copolymers that are obtained by copolymerization of two monomer species are sometimes termed bipolymers, those obtained from three monomers terpolymers, those obtained from four monomers quaterpolymers, etc.

Alternating copolymers: A copolymer consisting of macromolecules comprising two species of monomeric units in alternating sequence.

Note: An alternating copolymer may be considered as a homopolymer derived from an implicit or hypothetical monomer.

Block copolymers: A portion of a macromolecule, comprising many constitutional units, that has at least one feature which is not present in the adjacent portions.

Graft macromolecule: A macromolecule with one or more species of block connected to the main chain as side-chains, these side-chains having constitutional or configurational features that differ from those in the main chain.

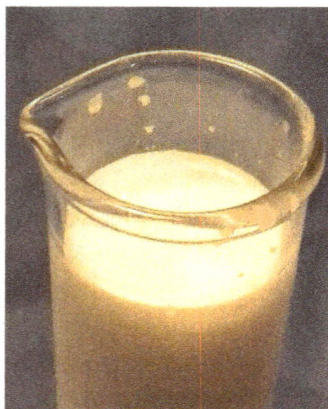

Vinyl Copolymer Milk

When two or more different monomers unite together to polymerize, their result is called a copolymer and its process is called copolymerization.

Commercially relevant copolymers include acrylonitrile butadiene styrene (ABS), styrene/butadiene co-polymer (SBR), nitrile rubber, styrene-acrylonitrile, styrene-isoprene-styrene (SIS) and ethylene-vinyl acetate.

Types of Copolymers

Since a copolymer consists of at least two types of constituent units (also structural units), copolymers can be classified based on how these units are arranged along the chain. These include:

```
—A—A—A—A—A—A—A—A—A—A—        1

—A—B—A—B—A—B—A—B—A—B—        2

—A—B—B—B—A—B—A—B—A—A—        3

—B—B—B—B—B—A—A—A—A—A—        4

—A—A—A—A—A—A—A—A—A—A—        5
     |                 |
—B—B—B          B—B—B —
```

Different types of copolymers

- Alternating copolymers with regular alternating A and B units (2)

- Periodic copolymers with A and B units arranged in a repeating sequence (e.g. $(A-B-A-B-B-A-A-A-A-B-B-B)_n$)

- Statistical copolymers are copolymers in which the sequence of monomer residues follows a statistical rule. If the probability of finding a given type monomer residue at a particular point in the chain is equal to the mole fraction of that monomer residue in the chain, then the polymer may be referred to as a truly random copolymer (3).

- Block copolymers comprise two or more homopolymer subunits linked by covalent bonds (4). The union of the homopolymer subunits may require an intermediate non-repeating subunit, known as a junction block. Block copolymers with two or three distinct blocks are called diblock copolymers and triblock copolymers, respectively.

Copolymers may also be described in terms of the existence of or arrangement of branches in the polymer structure. Linear copolymers consist of a single main chain whereas branched copolymers consist of a single main chain with one or more polymeric side chains.

Other special types of branched copolymers include star copolymers, brush copolymers, and comb copolymers. In gradient copolymers the monomer composition changes gradually along the chain.

A terpolymer is a copolymer consisting of three distinct monomers. The term is derived from *ter* (Latin), meaning thrice, and polymer.

- Stereoblock copolymers

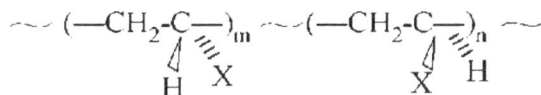

$$\sim\sim (-CH_2\text{-}C-)_m \sim\sim(-CH_2\text{-}C-)_n \sim\sim$$

A special structure can be formed from one monomer where now the distinguishing feature is the tacticity of each block.

Graft Copolymers

Graft copolymers are a special type of branched copolymer in which the side chains are structurally distinct from the main chain. The illustration (5) depicts a special case where the main chain and side chains are composed of distinct homopolymers. However, the individual chains of a graft copolymer may be homopolymers or copolymers. Note that different copolymer sequencing is sufficient to define a structural difference, thus an A-B diblock copolymer with A-B alternating copolymer side chains is properly called a graft copolymer.

For example, suppose we perform a free-radical polymerization of styrene in the presence of polybutadiene, a synthetic rubber, which retains one reactive C=C double bond per residue. We get polystyrene chains growing out in either direction from some of the places where there were double bonds, with a one-carbon rearrangement. Or to look at it the other way around, the result is a polystyrene backbone with polybutadiene chains growing out of it in both directions. This is an interesting copolymer variant in that one of the ingredients was a polymer to begin with.

As with block copolymers, the quasi-composite product has properties of both "components". In the example cited, the rubbery chains absorb energy when the substance is hit, so it is much less brittle than ordinary polystyrene. The product is called high-impact polystyrene, or HIPS.

Block Copolymers

One kind of copolymer is called a "block copolymer". Block copolymers are made up of blocks of different polymerized monomers and is usually made by first polymerizing styrene, and then subsequently polymerizing methyl methacrylate (MMA) from the reactive end of the polystyrene chains. This polymer is a "diblock copolymer" because it contains two different chemical blocks. Triblocks, tetrablocks, multiblocks, etc. can also be made. Diblock copolymers are made using living polymerization techniques, such as atom transfer free radical polymerization (ATRP), reversible addition fragmentation chain transfer (RAFT), ring-opening metathesis polymerization (ROMP), and living cationic or living anionic polymerizations. An emerging technique is chain shuttling polymerization.

The "blockiness" of a copolymer is a measure of the adjacency of comonomers vs their statistical distribution. Many or even most synthetic polymers are in fact copolymers, containing about 1-20% of a minority monomoner. In such cases, blockiness is undesirable.

Phase Separation

Block copolymers are interesting because they can "microphase separate" to form periodic nanostructures, as in the styrene-butadiene-styrene block copolymer shown at right. The polymer is known as Kraton and is used for shoe soles and adhesives. Owing to the microfine structure, the transmission electron microscope or TEM was needed to examine the structure. The butadiene matrix was stained with osmium tetroxide to provide contrast in the image. The material was made by living polymerization so that the blocks are almost monodisperse, so helping to create a very regular microstructure. The molecular weight of the polystyrene blocks in the main picture is 102,000; the inset picture has a molecular weight of 91,000, producing slightly smaller domains.

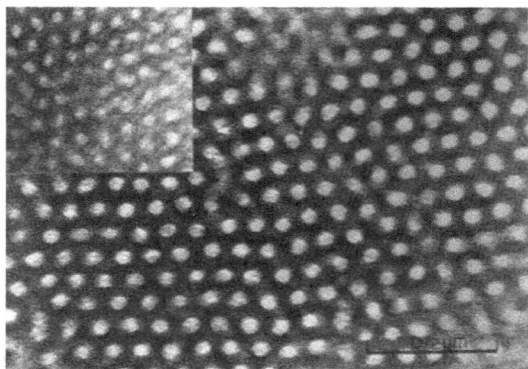

SBS block copolymer in TEM

Microphase separation is a situation similar to that of oil and water. Oil and water are immiscible - they phase separate. Due to incompatibility between the blocks, block copolymers undergo a similar phase separation. Because the blocks are covalently bonded to each other, they cannot demix macroscopically as water and oil. In "microphase separation" the blocks form nanometer-sized structures. Depending on the relative lengths of each block, several morphologies can be obtained. In diblock copolymers, sufficiently different block lengths lead to nanometer-sized spheres of one block in a matrix of the second (for example PMMA in polystyrene). Using less different block lengths, a "hexagonally packed cylinder" geometry can be obtained. Blocks of similar length form layers (often called lamellae in the technical literature). Between the cylindrical and lamellar phase is the gyroid phase. The nanoscale structures created from block copolymers could potentially be used for creating devices for use in computer memory, nanoscale-templating and nanoscale separations.

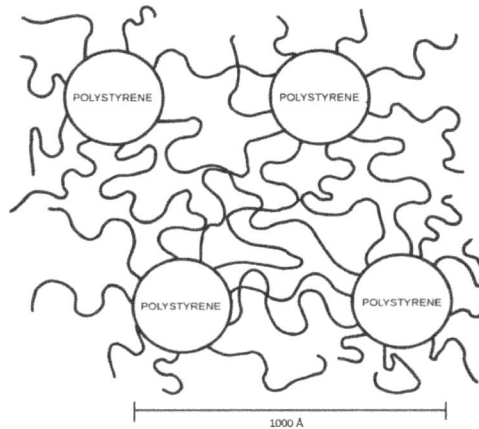

SBS block copolymer schematic microstructure

Polymer scientists use thermodynamics to describe how the different blocks interact. The product of the degree of polymerization, n, and the Flory-Huggins interaction parameter, χ, gives an indication of how incompatible the two blocks are and whether or not they will microphase separate. For example, a diblock copolymer of symmetric composition will microphase separate if the product χ is greater than 10.5. If χN is less than 10.5, the blocks will mix and microphase separation is not observed. The incompatibility between the blocks also affects the solution behavior of these copolymers and their adsorption behavior on various surfaces.

Copolymer Equation

An alternating copolymer has the formula: -A-B-A-B-A-B-A-B-A-B-, or -(-A-B-)$_n$-. The molar ratios of the monomer in the polymer is close to one, which happens when the reactivity ratios r$_1$ & r$_2$ are close to zero, as given by the Mayo–Lewis equation also called the copolymerization equation:

$$\frac{d[M_1]}{d[M_2]} = \frac{[M_1]\left(r_1[M_1]+[M_2]\right)}{[M_2]\left([M_1]+r_2[M_2]\right)}$$

where r$_1$ = k$_{11}$/k$_{12}$ & r$_2$ = k$_{22}$/k$_{21}$

Copolymer Engineering

Copolymerization is used to modify the properties of manufactured plastics to meet specific needs, for example to reduce crystallinity, modify glass transition temperature or to improve solubility. It is a way of improving mechanical properties, in a technique known as rubber toughening. Elastomeric phases within a rigid matrix act as crack arrestors, and so increase the energy absorption when the material is impacted for example. Acrylonitrile butadiene styrene is a common example.

Polyvinyl Chloride

Polyvinyl chloride, more correctly but unusually poly(vinyl chloride), commonly abbreviated PVC, is the third-most widely produced synthetic plastic polymer, after polyethylene and polypropylene.

PVC comes in two basic forms: rigid (sometimes abbreviated as RPVC) and flexible. The rigid form of PVC is used in construction for pipe and in profile applications such as doors and windows. It is also used for bottles, other non-food packaging, and cards (such as bank or membership cards). It can be made softer and more flexible by the addition of plasticizers, the most widely used being phthalates. In this form, it is also used in plumbing, electrical cable insulation, imitation leather, signage, inflatable products, and many applications where it replaces rubber.

Pure poly (vinyl chloride) is a white, brittle solid. It is insoluble in alcohol but slightly soluble in tetrahydrofuran.

Discovery

PVC was accidentally synthesized in 1872 by German chemist Eugen Baumann. The polymer appeared as a white solid inside a flask of vinyl chloride that had been left exposed to sunlight. In the early 20th century the Russian chemist Ivan Ostromislensky and Fritz Klatte of the German chemical company Griesheim-Elektron both attempted to use PVC in commercial products, but difficulties in processing the rigid, sometimes brittle polymer thwarted their efforts. Waldo Semon and the B.F. Goodrich Company developed a method in 1926 to plasticize PVC by blending it with various additives. The result was a more flexible and more easily processed material that soon achieved widespread commercial use.

Production

Polyvinyl chloride is produced by polymerization of the vinyl chloride monomer (VCM), as shown.

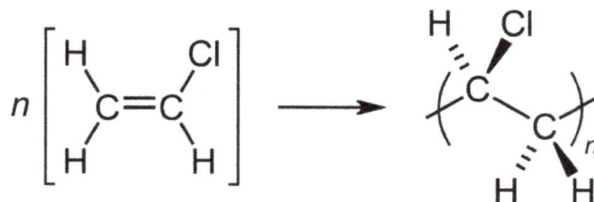

About 80% of production involves suspension polymerization. Emulsion polymerization accounts for about 12% and bulk polymerization accounts for 8%. Suspension polymerizations affords particles with average diameters of 100–180 μm, whereas emulsion polymerization gives much smaller particles of average size around 0.2 μm. VCM and water are introduced into the reactor and a polymerization initiator, along with other additives. The reaction vessel is pressure tight to contain the VCM. The contents of the reaction vessel are continually mixed to maintain the suspension and ensure a uniform particle size of the PVC resin. The reaction is exothermic, and thus requires cooling. As the volume is reduced during the reaction (PVC is denser than VCM), water is continually added to the mixture to maintain the suspension.

The polymerization of VCM is started by compounds called initiators that are mixed into the droplets. These compounds break down to start the radical chain reaction. Typical initiators include dioctanoyl peroxide and dicetyl peroxydicarbonate, both of which have fragile O-O bonds. Some initiators start the reaction rapidly but decay quickly and other initiators have the opposite effect. A combination of two different initiators is often used to give a uniform rate of polymerization. After the polymer has grown by about 10x, the short polymer precipitates inside the droplet of VCM, and polymerization continues with the precipitated, solvent-swollen particles. The weight average molecular weights of commercial polymers range from 100,000 to 200,000 and the number average molecular weights range from 45,000 to 64,000.

Once the reaction has run its course, the resulting PVC slurry is degassed and stripped to remove excess VCM, which is recycled. The polymer is then passed through a centrifuge to remove water. The slurry is further dried in a hot air bed, and the resulting powder sieved before storage or pelletization. Normally, the resulting PVC has a VCM content of less than 1 part per million. Other production processes, such as micro-suspension polymerization and emulsion polymerization, produce PVC with smaller particle sizes (10 μm vs. 120–150 μm for suspension PVC) with slightly different properties and with somewhat different sets of applications.

Microstructure

The polymers are linear and are strong. The monomers are mainly arranged head-to-tail, meaning that there are chlorides on alternating carbon centres. PVC has mainly an atactic stereochemistry, which means that the relative stereochemistry of the chloride centres are random. Some degree of syndiotacticity of the chain gives a few percent crystallinity that is influential on the properties of the material. About 57% of the mass of PVC is chlorine. The presence of chloride groups gives the polymer very different properties from the structurally related material polyethylene.

Additives to Finished Polymer

The product of the polymerization process is unmodified PVC. Before PVC can be made into finished products, it always requires conversion into a compound by the incorporation of additives (but not necessarily all of the following) such as heat stabilizers, UV stabilizers, plasticizers, processing aids, impact modifiers, thermal modifiers, fillers, flame retardants, biocides, blowing agents and smoke suppressors, and, optionally, pigments. The choice of additives used for the PVC finished product is controlled by the cost performance requirements of the end use specification e.g. underground pipe, window frames, intravenous tubing and flooring all have very different ingredients to suit their performance requirements. Previously, polychlorinated biphenyls (PCBs) were added to certain PVC products as flame retardants and stabilizers.

Phthalate Plasticizers

Most vinyl products contain plasticizers which dramatically improve their performance characteristic. The most common plasticizers are derivatives of phthalic acid. The materials are selected on their compatibility with the polymer, low volatility levels, and cost. These materials are usually oily colourless substances that mix well with the PVC particles. About 90% of the plasticizer market, estimated to be millions of tons per year worldwide, is dedicated to PVC.

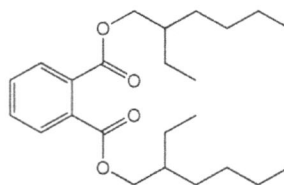

Bis(2-ethylhexyl) phthalate was a common plasticizer for PVC but is being replaced by higher molecular weight phthalates.

High and Low Molecular Weight Phthalates

Phthalates can be divided into three groups based on their molecular weight.

Low molecular weight phthalates have 6 or 7 carbon atoms in their alcohol chain. Medium molecular weight phthalates have 8 or 9 carbons in their alcohol chain. High molecular weight phthalates have from 10 to 13 carbons in their alcohol chain. Low molecular weight phthalates are no longer used in the US because of high volatility at pvc processing temperatures. The most common medium molecular weight phthalates are DOP (dioctyl phthalate, also known as DEHP, di-2-ethylhexyl phthalate) and DINP (diisononyl phthalate). High molecular weight phthalates have a limit of 12 carbons (if linear) or 13 carbons (if branched) in the alcohol chain because phthalates with longer alcohol chains are incompatible with pvc resin. The more common high molecular weight phthalates are DIDP, DPHP and DTDP, all from branched alcohols. There are phthalates from linear alcohols from 8 to 12 carbons which are used when the flexible pvc compound is to have superior low temperature flexibility and improved weathering. Because of possible health effects, there are movements to replace some phthalates such as DOP/DEHP with safer alternatives in Canada, the European Union, and the United States. In Europe DEHP and DBP are undergoing the REACH Authorisation process Registration, Evaluation, Authorisation and Restriction of Chemicals with results expected in 2014. Since no applications for BBP and DIBP were received by the European Chemicals Agency (ECHA), the use of these substances will be phased out in the EU by 21 February 2015.

Mid-high molecular weight phthalates today represent over 85% of all the phthalates currently being produced in Europe (2011). On 31 January 2014 the European Commission published its conclusions regarding the re-evaluation of the restrictions on DINP and DIDP in toys and childcare articles which can be placed in the mouth.The Commission confirmed the main conclusions presented in August last year by the ECHA, which was asked in September 2009 to review any newly available scientific information relative to these two high phthalates.

The European Commission concluded that "no unacceptable risk has been characterised for the uses of DINP and DIDP in articles other than toys and childcare articles which can be placed in the mouth". With regard to the latter, the existent restrictions are nevertheless maintained, based on the precautionary principle. DINP and DIDP are therefore safe for use in all current consumer applications. The European Chemicals Agency also concluded that no further risk management measures are needed to reduce the exposure of adults and children to DINP and DIDP.

Heat Stabilizers

One of the most crucial additives are heat stabilizers. These agents minimize loss of HCl, a degradation process that starts above 70 °C. Once dehydrochlorination starts, it is autocatalytic. Many diverse agents have been used including, traditionally, derivatives of heavy metals (lead, cadmium). Increasingly, metallic soaps (metal "salts" of fatty acids) are favored, species such as calcium stearate. Addition levels vary typically from 2% to 4%. The choice of the best heat stabilizer depends on its cost effectiveness in the end use application, performance specification requirements, processing technology and regulatory approvals.

Rigid PVC Applications

Regular PVC (polyvinyl chloride) is a common, strong but lightweight plastic used in construction. It is made softer and more flexible by the addition of plasticizers. If no plasticizers are added, it is known as uPVC (unplasticized polyvinyl chloride), rigid PVC, or vinyl siding in the U.S. In Europe, particularly Belgium, there has been a commitment to eliminate the use of cadmium (previously used as a part component of heat stabilizers in window profiles) and phase out lead based heat stabilizers (as used in pipe and profile areas) such as liquid autodiachromate and calcium polyhydrocummate by 2015. According to the final report of Vinyl 2010 cadmium was eliminated across Europe by 2007. The progressive substitution of lead-based stabilizers is also confirmed in the same document showing a reduction of 75% since 2000 and ongoing. This is confirmed by the corresponding growth in calcium-based stabilizers, used as an alternative to lead-based stabilizers, more and more, also outside Europe.

Tin based stabilizers are mainly used in Europe for rigid, transparent applications due to the high temperature processing conditions used. The situation in North America is different where tin systems are used for almost all rigid PVC applications. Tin stabilizers can be divided into two main groups, the first group containing those with tin-oxygen bonds and the second group with tin-sulphur bonds. According to the European Stabiliser producers most organotin stabilisers have already been successfully REACH registered. More chemical and use information is also available on this site.

Flexible PVC Applications

Flexible PVC coated wire and cable for electrical use has traditionally been stabilised with lead but these are being replaced, as in the rigid area, with calcium based systems.

Liquid mixed metal stabilisers are used in several PVC flexible applications such as calendered films, extruded profiles, injection moulded soles and footwear, extruded hoses and plastisols where PVC paste is spread on to a backing (flooring, wall covering, artificial leather). Liquid mixed metal stabiliser systems are primarily based on barium, zinc and calcium carboxylates. In general liquid mixed metals like BaZn, CaZn require the addition of co-stabilisers, antioxidants and organo-phosphites to provide optimum performance.

BaZn stabilisers have successfully replaced cadmium-based stabilisers in Europe in many PVC semi-rigid and flexible applications according to the European producers.

Physical Properties

PVC is a thermoplastic polymer. Its properties are usually categorized based on rigid and flexible PVCs.

Property	Rigid PVC	Flexible PVC
Density [g/cm^3]	1.3–1.45	1.1–1.35
Thermal conductivity [W/(m·K)]	0.14–0.28	0.14–0.17
Yield strength [psi]	4500–8700	1450–3600
Young's modulus [psi]	490,000	
Flexural strength (yield) [psi]	10,500	
Compression strength [psi]	9500	
Coefficient of thermal expansion (linear) [mm/(mm °C)]	5×10^{-5}	
Vicat B [°C]	65–100	Not recommended
Resistivity [Ω m]	10^{16}	10^{12}–10^{15}
Surface resistivity [Ω]	10^{13}–10^{14}	10^{11}–10^{12}

Mechanical Properties

PVC has high hardness and mechanical properties. The mechanical properties enhance with the molecular weight increasing but decrease with the temperature increasing. The mechanical properties of rigid PVC (uPVC) are very good; the elastic modulus can reach 1500-3,000 MPa. The soft PVC (flexible PVC) elastic is 1.5–15 MPa.

Thermal and Fire Properties

The heat stability of raw PVC is very poor, so the addition of a heat stabilizer during the process is necessary in order to ensure the product's properties. PVC starts to decompose when the temperature reaches 140 °C, with melting temperature starting around 160 °C. The linear expansion coefficient of rigid PVC is small and has good flame retardancy, the Limiting oxygen index (LOI) being up to 45 or more. The LOI is the minimum concentration of oxygen, expressed as a percentage, that will support combustion of a polymer and noting that air has 20% content of oxygen.

Electrical Properties

PVC is a polymer with good insulation properties, but because of its higher polar nature the electrical insulating property is inferior to non polar polymers such as polyethylene and polypropylene.

Since the dielectric constant, dielectric loss tangent value, and volume resistivity are high, the corona resistance is not very good, and it is generally suitable for medium or low voltage and low frequency insulation materials.

Chemical Properties

PVC is chemical resistant to acids, salts, bases, fats, and alcohols therefore it is used in sewerage piping. It is also resistant to some solvents, mainly uPVC. Plasticized PVC is in some cases less resistant to solvents. For example, PVC is resistant to fuel and some paint thinners. Some solvents may only swell it or deform it but not dissolve it, but some of them, like tetrahydrofuran or acetone, may damage it.

Applications

PVC's relatively low cost, biological and chemical resistance and workability have resulted in it being used for a wide variety of applications. It is used for sewerage pipes and other pipe applications where cost or vulnerability to corrosion limit the use of metal. With the addition of impact modifiers and stabilizers, PVC scrap has become a popular material for window and door which is 50% less than the cost of wooden window and door. By adding plasticizers, it can become flexible enough to be used in cabling applications as a wire insulator. It has been used in many other applications. In 2013, about 39.3 million tonnes of PVC were consumed worldwide. PVC demand is forecast to increase at an average annual rate of 3.2% until 2021.

PVC is used extensively in sewage pipe due to its low cost, chemical resistance and ease of jointing

Pipes

Roughly half of the world's polyvinyl chloride resin manufactured annually is used for producing pipes for municipal and industrial applications. In the water distribution market it accounts for 66% of the market in the US, and in sanitary sewer pipe applications, it accounts for 75%. Its light weight, low cost, and low maintenance make it attractive. However, it must be carefully installed and bedded to ensure longitudinal cracking and overbelling does not occur. Additionally, PVC pipes can be fused together using various solvent cements, or heat-fused (butt-fusion process, similar to joining HDPE pipe), creating permanent joints that are virtually impervious to leakage.

In February 2007 the California Building Standards Code was updated to approve the use of chlorinated polyvinyl chloride (CPVC) pipe for use in residential water supply piping systems. CPVC has been a nationally accepted material in the US since 1982; California, however, has permitted only limited use since 2001. The Department of Housing and Community Development prepared and certified an environmental impact statement resulting in a recommendation that the Commission adopt and approve the use of CPVC. The Commission's vote was unanimous and CPVC has been placed in the 2007 California Plumbing Code.

In the United States and Canada, PVC pipes account for the largest majority of pipe materials used in buried municipal applications for drinking water distribution and wastewater mains. Buried PVC pipes in both water and sanitary sewer applications that are 4 inches (100 mm) in diameter and larger are typically joined by means of a gasket-sealed joint. The most common type of gasket utilized in North America is a metal reinforced elastomer, commonly referred to as a Rieber sealing system.

Electric Cables

PVC is commonly used as the insulation on electrical cables; PVC used for this purpose needs to be plasticized.

In a fire, PVC-coated wires can form hydrogen chloride fumes; the chlorine serves to scavenge free radicals and is the source of the material's fire retardance. While HCl fumes can also pose a health hazard in their own right, HCl dissolves in moisture and breaks down onto surfaces, particularly in areas where the air is cool enough to breathe, and is not available for inhalation. Frequently in applications where smoke is a major hazard (notably in tunnels and communal areas) PVC-free cable insulation is preferred, such as low smoke zero halogen (LSZH) insulation.

Unplasticized Poly(Vinyl Chloride) (uPVC) for Construction

uPVC, also known as rigid PVC, is extensively used in the building industry as a low-maintenance material, particularly in Ireland, the United Kingdom, in the United States and Canada. In the USA and Canada it is known as vinyl, or vinyl siding. The material comes in a range of colors and finishes, including a photo-effect wood finish, and is used as a substitute for painted wood, mostly for window frames and sills when installing double glazing in new buildings, or to replace older single-glazed windows. Other uses include fascia, and siding or weatherboarding. This material has almost entirely replaced the use of cast iron for plumbing and drainage, being used for waste pipes, drainpipes, gutters and downspouts. uPVC does not contain phthalates, since those are only added to flexible PVC, nor does it contain BPA. uPVC is known as having strong resistance against chemicals, sunlight, and oxidation from water.

"A modern Tudorbethan" house with uPVC gutters and downspouts, fascia, decorative imitation "half-timbering", windows, and doors

Double glazed units

Signs

Poly(vinyl chloride) is formed in flat sheets in a variety of thicknesses and colors. As flat sheets, PVC is often expanded to create voids in the interior of the material, providing additional thickness without additional weight and minimal extra cost (Closed-cell PVC foamboard). Sheets are cut using saw and rotary cutting equipment. Plasticized PVC is also used to produce thin, colored, or clear, adhesive-backed films referred to simply as vinyl. These films are typically cut on a computer-controlled plotter or printed in a wide-format printer. These sheets and films are used to produce a wide variety of commercial signage products and markings on vehicles, e.g. car body stripes.

Clothing and Furniture

PVC has become widely used in clothing, to either create a leather-like material or at times simply for the effect of PVC. PVC clothing is common in Goth, Punk, clothing fetish and alternative fashions. PVC is less expensive than rubber, leather, and latex which it is therefore used to simulate.

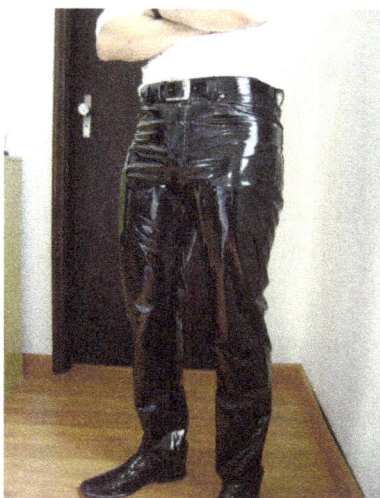

Black PVC trousers

PVC fabric is water-resistant so is used in coats, skiing equipment, shoes, jackets, aprons, and bags.

Healthcare

The two main application areas for single use medically approved PVC compounds are flexible containers and tubing: containers used for blood and blood components, for urine collection or for ostomy products and tubing used for blood taking and blood giving sets, catheters, heart-lung bypass sets, hemodialysis sets etc. In Europe the consumption of PVC for medical devices is approximately 85.000 tons every year. Almost one third of plastic based medical devices are made from PVC. The reasons for using flexible PVC in these applications for over 50 years are numerous and based on cost effectiveness linked to transparency, light weight, softness, tear strength, kink resistance, suitability for sterilization and biocompatibility.

Plasticisers

DEHP (Di-2ethylhexylphthalate) has been medically approved for many years for use in such medical devices; the PVC-DEHP combination proving to be very suitable for making blood bags because DEHP stabilises red blood cells, minimising haemolysis (red blood cell rupture). However, DEHP is coming under increasing pressure in Europe. The assessment of potential risks related to phthalates, and in particular the use of DEHP in PVC medical devices, was subject to scientific and policy review by the European Union authorities, and on 21 March 2010, a specific labelling requirement was introduced across the EU for all devices containing phthalates that are classified as CMR (carcinogenic, mutagenic or toxic to reproduction). The label aims to enable healthcare professionals to use this equipment safely, and, where needed, take appropriate precautionary measures for patients at risk of over-exposure.

DEHP alternatives, which are gradually replacing it, are Adipates, Butyryltrihexylcitrate (BTHC), Cyclohexane-1,2-dicarboxylic acid, diisononylester (DINCH), Di(2-ethylhexyl)terephthalate, polymerics and trimellitic acid, 2-ethylhexylester (TOTM).

Flooring

Flexible PVC flooring is inexpensive and used in a variety of buildings covering the home, hospitals, offices, schools, etc. Complex and 3D designs are possible due to the prints that can be created which are then protected by a clear wear layer. A middle vinyl foam layer also gives a comfortable and safe feel. The smooth, tough surface of the upper wear layer prevents the buildup of dirt, which prevents microbes from breeding in areas that need to be kept sterile, such as hospitals and clinics.

Other Applications

PVC has been used for a host of consumer products of relatively smaller volume compared to the industrial and commercial applications described above. Another of its earliest mass-market consumer applications was to make vinyl records. More recent examples include wallcovering, greenhouses, home playgrounds, foam and other toys, custom truck toppers (tarpaulins), ceiling tiles and other kinds of interior cladding.

PVC piping is cheaper than metals used in musical instrument making; it is therefore a common alternative when making instruments, often for leisure or for rarer instruments such as the contrabass flute.

The handles of the Victorinox Swiss Army knives are made of PVC.

Chlorinated PVC

PVC can be usefully modified by chlorination, which increases its chlorine content to 67%. The new material has a higher heat resistance so is primarily used for hot water pipe and fittings, but it is more expensive and it is found only in niche applications, such as certain water heaters and certain specialized clothing. An extensive market for chlorinated PVC is in pipe for use in office building, apartment and condominium fire protection. CPVC, as it is called, is produced by chlorination of aqueous solution of suspension PVC particles followed by exposure to UV light which initiates the free-radical chlorination.

Health and Safety

Degradation

Degradation during service life, or after careless disposal, is a chemical change that drastically reduces the average molecular weight of the polyvinyl chloride polymer. Since the mechanical integrity of a plastic depends on its high average molecular weight, wear and tear inevitably weakens the material. Weathering degradation of plastics results in their surface embrittlement and microcracking, yielding microparticles that continue on in the environment. Also known as microplastics, these particles act like sponges and soak up Persistent Organic Pollutants (POPs) around them. Thus laden with high levels of POPs, the microparticles are often ingested by organisms in the biosphere.

However, there is evidence that three of the polymers (HDPE, LDPE, and PP) consistently soaked up POPs at concentrations an order of magnitude higher than did the remaining two (PVC and PET). After 12 months of exposure, for example, there was a 34-fold difference in average total POPs amassed on LDPE compared to PET at one location. At another site, average total POPs adhered to HDPE was nearly 30 times that of PVC. The researchers think that differences in the size and shape of the polymer molecules can explain why some accumulate more pollutants than others. The fungus Aspergillus fumigatus effectively degrades plasticized PVC. Phanerochaete chrysosporium was grown on PVC in a mineral salt agar. Phanerochaete chrysosporium, Lentinus tigrinus, Aspergillus niger, and Aspergillus sydowii can effectively degrade PVC.

Plasticizers

Phthalates, which are incorporated into plastics as plasticizers comprise 70% of the U.S. plasticizer market; phthalates are by design not covalently bound to the polymer matrix, which makes them highly susceptible to leaching. Phthalates are contained in plastics at high percentages. For example, they can contribute up to 40% by weight to intravenous medical bags and up to 80% by weight in medical tubing. Vinyl products are pervasive—including toys, car interiors, shower curtains, and flooring—and initially release chemical gases into the air. Some studies indicate that this outgassing of additives may contribute to health complications, and have resulted in a call for banning the use of DEHP on shower curtains, among other uses. The Japanese car companies Toyota, Nissan, and Honda have eliminated PVC in their car interiors starting in 2007.

In 2004 a joint Swedish-Danish research team found a statistical association between allergies in children and indoor air levels of DEHP and BBzP (butyl benzyl phthalate), which is used in vinyl flooring. In December 2006, the European Chemicals Bureau of the European Commission released a final draft risk assessment of BBzP which found "no concern" for consumer exposure including exposure to children.

EU Decisions on Phthalates

Risk assessments have led to the classification of low molecular weight and labeling as Category 1B Reproductive agents. Three of these phthalates, DBP, BBP and DEHP were included on annex XIV of the REACH regulation in February 2011 and will be phased out by the EU by February 2015 unless an application for authorisation is made before July 2013 and an authorisation granted. DIBP is still on the REACH Candidate List for Authorisation. Environmental Science & Technology, a peer reviewed journal published by the American Chemical Society states DEHP poses a serious risk to human health.

In 2008 the European Union's Scientific Committee on Emerging and Newly Identified Health Risks (SCENIHR) reviewed the safety of DEHP in medical devices. The SCENIHR report states that certain medical procedures used in high risk patients result in a significant exposure to DEHP and concludes there is still a reason for having some concerns about the exposure of prematurely born male babies to medical devices containing DEHP. The Committee said there are some alternative plasticizers available for which there is sufficient toxicological data to indicate a lower hazard compared to DEHP but added that the functionality of these plasticizers should be assessed before they can be used as an alternative for DEHP in PVC medical devices. Risk assessment results have shown positive results regarding the safe use of High Molecular Weight Phthalates. They have all been registered for REACH and do not require any classification for health and environmental effects, nor are they on the Candidate List for Authorisation. High phthalates are not CMR (carcinogenic, mutagenic or toxic for reproduction), neither are they considered endocrine disruptors.

In the EU Risk Assessment the European Commission has confirmed that Di-isononyl phthalate (DINP) and Di-isodecyl phthalate (DIDP) pose no risk to either human health or the environment from any current use. The European Commission's findings (published in the EU Official Journal on 13 April 2006) confirm the outcome of a risk assessment involving more than 10 years of extensive scientific evaluation by EU regulators. Following the recent adoption of EU legislation with the regard to the marketing and use of DINP in toys and childcare articles, the risk assessment conclusions clearly state that there is no need for any further measures to regulate the use of DINP. In Europe and in some other parts of the world, the use of DINP in toys and childcare items has been restricted as a precautionary measure. In Europe, for example, DINP can no longer be used in toys and childcare items that can be put in the mouth even though the EU scientific risk assessment concluded that its use in toys does not pose a risk to human health or the environment. The rigorous EU risk assessments, which include a high degree of conservatism and built-in safety factors, have been carried out under the strict supervision of the European Commission and provide a clear scientific evaluation on which to judge whether or not a particular substance can be safely used.

The FDA Paper titled "Safety Assessment of Di(2-ethylhexyl)phthalate (DEHP)Released from PVC Medical Devices" states that [3.2.1.3] Critically ill or injured patients may be at increased risk of developing adverse health effects from DEHP, not only by virtue of increased exposure relative to the general population, but also because of the physiological and pharmacodynamic changes that occur in these patients compared to healthy individuals.

Lead

The metal lead had previously been frequently added to PVC to improve workability and stability. Lead has been shown to leach into drinking water from PVC pipes.

In Europe (EU 27) the use of Lead stabilizers is being gradually phased out by 2015 under the VinylPlus voluntary commitment with over 80% replaced by 2013.

Vinyl Chloride Monomer

In the early 1970s, the carcinogenicity of vinyl chloride (usually called vinyl chloride monomer or VCM) was linked to cancers in workers in the polyvinyl chloride industry. Specifically workers in polymerization section of a B.F. Goodrich plant near Louisville, Kentucky (US) were diagnosed with liver angiosarcoma also known as hemangiosarcoma, a rare disease. Since that time, studies of PVC workers in Australia, Italy, Germany, and the UK have all associated certain types of occupational cancers with exposure to vinyl chloride, and it has become accepted that VCM is a carcinogen. Technology for removal of VCM from products has become stringent, commensurate with the associated regulations.

Dioxins

PVC produces HCl upon combustion almost quantitatively related to its chlorine content. Extensive studies in Europe indicate that the chlorine found in emitted dioxins is not derived from HCl in the flue gases. Instead, most dioxins arise in the condensed solid phase by the reaction of inorganic chlorides with graphitic structures in char-containing ash particles. Copper acts as a catalyst for these reactions.

Studies of household waste burning indicate consistent increases in dioxin generation with increasing PVC concentrations. According to the EPA dioxin inventory, landfill fires are likely to represent an even larger source of dioxin to the environment. A survey of international studies consistently identifies high dioxin concentrations in areas affected by open waste burning and a study that looked at the homologue pattern found the sample with the highest dioxin concentration was "typical for the pyrolysis of PVC". Other EU studies indicate that PVC likely "accounts for the overwhelming majority of chlorine that is available for dioxin formation during landfill fires."

The next largest sources of dioxin in the EPA inventory are medical and municipal waste incinerators. Various studies have been conducted that reach contradictory results. For instance a study of commercial-scale incinerators showed no relationship between the PVC content of the waste and dioxin emissions. Other studies have shown a clear correlation between dioxin formation and chloride content and indicate that PVC is a significant contributor to the formation of both dioxin and PCB in incinerators.

In February 2007, the Technical and Scientific Advisory Committee of the US Green Building Council (USGBC) released its report on a PVC avoidance related materials credit for the LEED Green Building Rating system. The report concludes that "no single material shows up as the best across all the human health and environmental impact categories, nor as the worst" but that the "risk of dioxin emissions puts PVC consistently among the worst materials for human health impacts."

In Europe the overwhelming importance of combustion conditions on dioxin formation has been established by numerous researchers. The single most important factor in forming dioxin-like compounds is the temperature of the combustion gases. Oxygen concentration also plays a major role on dioxin formation, but not the chlorine content.

The design of modern incinerators minimises PCDD/F formation by optimising the stability of the thermal process. To comply with the EU emission limit of 0.1 ng I-TEQ/m3 modern incinerators operate in conditions minimising dioxin formation and are equipped with pollution control devices which catch the low amounts produced. Recent information is showing for example that dioxin levels in populations near incinerators in Lisbon and Madeira have not risen since the plants began operating in 1999 and 2002 respectively.

Several studies have also shown that removing PVC from waste would not significantly reduce the quantity of dioxins emitted. The European Union Commission published in July 2000 a Green Paper on the Environmental Issues of PVC. " The Commission states (page 27) that it has been suggested that the reduction of the chlorine content in the waste can contribute to the reduction of dioxin formation, even though the actual mechanism is not fully understood. The influence on the reduction is also expected to be a second or third order relationship. It is most likely that the main incineration parameters, such as the temperature and the oxygen concentration, have a major influence on the dioxin formation". The Green Paper states further that at the current levels of chlorine in municipal waste, there does not seem to be a direct quantitative relationship between chlorine content and dioxin formation.

A study commissioned by the European Commission on "Life Cycle Assessment of PVC and of principal competing materials" states that "Recent studies show that the presence of PVC has no significant effect on the amount of dioxins released through incineration of plastic waste."

End-of-life

The European waste hierarchy refers to the five steps included in the article 4 of the Waste Framework Directive:

1. Prevention: preventing and reducing waste generation.

2. Reuse and preparation for reuse: giving the products a second life before they become waste.

3. Recycle: any recovery operation by which waste materials are reprocessed into products, materials or substances whether for the original or other purposes. It includes composting and it does not include incineration.

4. Recovery: some waste incineration based on a political non-scientific formulathat upgrades the less inefficient incinerators.

5. Disposal: processes to dispose of waste be it landfilling, incineration, pyrolysis, gasification and other finalist solutions. Landfill is restricted in some EU-countries through Landfill Directives and there is a debate about Incineration E.g. original plastic which contains a lot of energy is just recovered in energy and not recycled. According to the Waste Framework Directive the European Waste Hierarchy is legally binding except in cases that may require

specific waste streams to depart from the hierarchy. This should be justified on the basis of life-cycle thinking.

The European Commission has set new rules to promote the recovery of PVC waste for use in a number of construction products. It says: "The use of recovered PVC should be encouraged in the manufacture of certain construction products because it allows the reuse of old PVC ... This avoids PVC being discarded in landfills or incinerated causing release of carbon dioxide and cadmium in the environment".

Industry Initiatives

In Europe, developments in PVC waste management have been monitored by Vinyl 2010, established in 2000. Vinyl 2010's objective was to recycle 200,000 tonnes of post-consumer PVC waste per year in Europe by the end of 2010, excluding waste streams already subject to other or more specific legislation (such as the European Directives on End-of-Life Vehicles, Packaging and Waste Electric and Electronic Equipment).

Since June 2011, it is followed by VinylPlus, a new set of targets for sustainable development. Its main target is to recycle 800,000 tonnes/year of PVC by 2020 including 100,000 tonnes of *difficult to recycle* waste. One facilitator for collection and recycling of PVC waste is Recovinyl. The reported and audited recycled PVC tonnage in 2014 was 481,018 tonnes

One approach to address the problem of waste PVC is also through the process called Vinyloop. It is a mechanical recycling process using a solvent to separate PVC from other materials. This solvent turns in a closed loop process in which the solvent is recycled. Recycled PVC is used in place of virgin PVC in various applications: coatings for swimming pools, shoe soles, hoses, diaphragms tunnel, coated fabrics, PVC sheets. This recycled PVC's primary energy demand is 46 percent lower than conventional produced PVC. So the use of recycled material leads to a significant better ecological footprint. The global warming potential is 39 percent lower.

Restrictions

In November 2005 one of the largest hospital networks in the U.S., Catholic Healthcare West, signed a contract with B. Braun Melsungen for vinyl-free intravenous bags and tubing.

In January 2012 a major U.S. West Coast healthcare provider, Kaiser Permanente, announced that it will no longer buy intravenous (IV) medical equipment made with polyvinyl chloride (PVC) and DEHP (di-2-ethyl hexyl phthalate) type plasticizers.

Sustainability

PVC is made from petroleum. The production process also uses sodium chloride. Recycled PVC is broken down into small chips, impurities removed, and the product refined to make pure white PVC. It can be recycled roughly seven times and has a lifespan of around 140 years.

In the UK, approximately 400 tonnes of PVC are recycled every month. Property owners can recycle it through nationwide collection depots. The Olympic Delivery Authority (ODA), for example, after initially rejecting PVC as material for different temporary venues of the London Olympics

2012, has reviewed its decision and developed a policy for its use. This policy highlighted that the functional properties of PVC make it the most appropriate material in certain circumstances while taking into consideration the environmental and social impacts across the whole life cycle, e.g. the rate for recycling or reuse and the percentage of recycled content. Temporary parts, like roofing covers of the Olympic Stadium, the Water Polo Arena, and the Royal Artillery Barracks, would be deconstructed and a part recycled in the Vinyloop process.

References

• Lee, Thomas H. (2004). The design of CMOS radio-frequency integrated circuits. Cambridge University Press. p. 20. ISBN 0-521-83539-9.

• Mueller, Gerd (2000) Electroluminescence I, Academic Press, ISBN 0-12-752173-9, p. 67, "escape cone of light" from semiconductor, illustrations of light cones on p. 69

• Elektrotechnik Gesamtband Technische Mathematik Kommunikationselektronik (in German) (1st ed.). Westermann. 1997. p. 171. ISBN 3142212515.

• Narra, Prathyusha; Zinger, D.S. (2004). "An effective LED dimming approach". Industry Applications Conference, 2004. 39th IAS Annual Meeting. Conference Record of the 2004 IEEE. 3: 1671–1676. doi:10.1109/IAS.2004.1348695. ISBN 0-7803-8486-5.

• Cox, James F. (2001). Fundamentals of linear electronics: integrated and discrete. Cengage Learning. pp. 91–. ISBN 978-0-7668-3018-9.

• Tavernier, Filip and Steyaert, Michiel (2011) High-Speed Optical Receivers with Integrated Photodiode in Nanoscale CMOS. Springer. ISBN 1-4419-9924-8.

• Riordan, Michael and Hoddeson, Lillian (1998). Crystal Fire: The Invention of the Transistor and the Birth of the Information Age. ISBN 9780393318517.

• Held. G, Introduction to Light Emitting Diode Technology and Applications, CRC Press, (Worldwide, 2008). Ch. 5 p. 116. ISBN 1-4200-7662-0

• Gao, Wei (2010). Precision Nanometrology: Sensors and Measuring Systems for Nanomanufacturing. Springer. pp. 15–16. ISBN 978-1-84996-253-7.

• Gevorkian, Peter (2007). Sustainable energy systems engineering: the complete green building design resource. McGraw Hill Professional. ISBN 978-0-07-147359-0.

• Tsokos, K. A. (28 January 2010). Physics for the IB Diploma Full Colour. Cambridge University Press. ISBN 978-0-521-13821-5.

• Pearce, J.; Lau, A. (2002). "Net Energy Analysis for Sustainable Energy Production from Silicon Based Solar Cells". Solar Energy (PDF). p. 181. doi:10.1115/SED2002-1051. ISBN 0-7918-1689-3.

• Yablonovitch, Eli; Miller, Owen D.; Kurtz, S. R. (2012). "The opto-electronic physics that broke the efficiency limit in solar cells". 2012 38th IEEE Photovoltaic Specialists Conference. p. 001556. doi:10.1109/PVSC.2012.6317891. ISBN 978-1-4673-0066-7.

• Suk-Joong, L. Kang Sintering -Densification, Grain Growth and Microstructure, Elsevier Butterworth-Heinemann, 2005, ISBN 0-7506-6385-5, p.3

• Richard G. Jones; Edward S. Wilks; W. Val Metanomski; Jaroslav Kahovec; Michael Hess; Robert Stepto; Tatsuki Kitayama, eds. (2009). Compendium of Polymer Terminology and Nomenclature (IUPAC Recommendations 2008) (2nd ed.). RSC Publ. p. 464. ISBN 978-0-85404-491-7.

• Lekkerkerker, Henk N.W.; Tuinier, Remco (2011). Colloids and the Depletion Interaction. Heidelberg: Springer. doi:10.1007/978-94-007-1223-2. ISBN 9789400712225.

• Menachem Elimelech; John Gregory; Xiadong Jia; Richard Williams (1998). Particle deposition and aggregation: measurement, modelling and simulation. Butterworth-Heinemann. ISBN 0-7506-7024-X.

- Salager, J-L (2000). Françoise Nielloud; Gilberte Marti-Mestres, eds. Pharmaceutical emulsions and suspensions. CRC press. p. 89. ISBN 0-8247-0304-9.

- Frimmel, Fritz H.; Frank von der Kammer; Hans-Curt Flemming (2007). Colloidal transport in porous media (1 ed.). Springer. p. 292. ISBN 3-540-71338-7.

- Terech P. (1997) "Low-molecular weight organogelators", pp. 208–268 in: Robb I.D. (ed.) Specialist surfactants. Glasgow: Blackie Academic and Professional, ISBN 0751403407.

Permissions

Index

www.ingramcontent.com/pod-product-compliance
Lightning Source LLC
Chambersburg PA
CBHW082021190326
41458CB00010B/3236